I0505547

II

III

Steve Halitsky

Edward Halitsky

THETA CRITERIA

Multivariate Analysis

Steve Halitsky
Edward Halitsky

www.ThetaCriteria.com

Mathematics Subject Classification:

46G12	Measures and integration on abstract linear spaces
58C40	Spectral theory; eigenvalue problems
58D30	Applications

BISAC:
Mathematics / Probability & Statistics / Multivariate Analysis

---

Halitsky, Steve
Halitsky, Edward
Theta Criteria: Multivariate Analysis

Bibliography: Separate to each Chapter
Includes Index

Copyright © Steve Halitsky and Edward Halitsky. All rights reserved.

No part of this book may be translated or reproduced in any form without specific written permission from the Authors.

Cover Photo: Øresundsbroen bridge from Sweden to Denmark.
Copyright © Roger Petersson.

MATLAB® is a registered trademark of MathWorks, Inc.
SAS® is a registered trademark of SAS Institute, Inc.
SPSS® is a registered trademark of SPSS, an IBM Company.

Dedicated to

Roxanne Halitsky

Introduction

In this book we will describe Theta Criteria properties and their application areas.

Theta Criteria are based on operator's theory and matrices decompositions. The main idea is to evaluate differences between sets of operators' eigenvalues & eigenvectors or eigenfunctions.

Theta Criteria will lead to more efficient analysis, optimization and prognosis of multivariate systems and applications. Theta Criteria can be used for data set mining and image processing. The book is designed as a complementary tool for applied mathematics methods.

We assume the Reader has standard undergraduate knowledge of advanced calculus, matrix theory, operator theory, approximation methods and measure theory.

The book consists of Introduction and four Chapters.

In Chapter I we briefly review existing methods and software packages for multivariate systems analysis. These methods are based upon Multivariate General Linear Hypothesis (MGLH).

In line with MGLH, all dataset variables are linear, additive and relationships models are linear series of weighted terms.

We also mention the following methods: Multiple Regression, Discriminant Function Analysis, Canonical Analysis, Principle

Components Analysis and formal linear algebra methods. Lastly, we discuss our methods, Theta Criteria, which are constructed on norms of weighted differences of matrices ordered eigenvectors.

In Chapter II, we placed our main formal results. At this moment we considered only positively-defined matrices.

In Chapter III we have numerically studied Theta Criteria on sets of varied matrices. We compare Theta Criteria accuracy with existing matrices' norms and invariants.

In Chapter IV, we briefly discuss Theta Criteria application areas.

At this moment we are only able to "scratch the surface" of Theta Criteria universe. We are convinced that our methods will find their places in various complex applications.

We express our gratitude to Academic Vladimir Skurichin (Ukrainian Academy of Science) for his guidance.

Notation

R is real numbers field

R^n -finite linear vector space over R

$R^{n\times n}$ -set of all positively defined matrices of order n

$\mathbf{R},\hat{\mathbf{R}}\in R^{n\times n}$; $\operatorname{rank}(\mathbf{R})=\operatorname{rank}(\hat{\mathbf{R}})=n$, $\mathbf{R}=\begin{bmatrix}\mathbf{A}_1 & \mathbf{B}^T\\ \mathbf{B} & \mathbf{A}_2\end{bmatrix}$, $\hat{\mathbf{R}}=\begin{bmatrix}\hat{\mathbf{A}}_1 & \hat{\mathbf{B}}^T\\ \hat{\mathbf{B}} & \hat{\mathbf{A}}_2\end{bmatrix}$ -

block matrices

$\mathbf{A}_1=[a_{ij}]_{i,j=\overline{1,m}}$, $\mathbf{A}_2=[a_{ij}]_{i,j=\overline{m+1,n}}$, $\mathbf{B}=[a_{ij}]_{i=\overline{m+1,n};j=\overline{1,m}}$ - sub matrices

$\hat{\mathbf{A}}_1=[\hat{a}_{ij}]_{i,j=\overline{1,m}}$, $\hat{\mathbf{A}}_2=[\hat{a}_{ij}]_{i,j=\overline{m+1,n}}$, $\hat{\mathbf{B}}=[\hat{a}_{ij}]_{i=\overline{m+1,n};j=\overline{1,m}}$ - sub matrices

$\Lambda=\{\lambda_i\}_{i=\overline{1,n}}$, $\hat{\Lambda}=\{\hat{\lambda}_i\}_{i=\overline{1,n}}$ - sets of all ordered eigenvalues of $\mathbf{R},\hat{\mathbf{R}}$,

$$\lambda_i>\lambda_j\ ,\quad \hat{\lambda}_i>\hat{\lambda}_j\ ,\quad i<j\ ,\quad \forall\ i,j=\overline{1,n}$$

$E=\{\mathbf{e}_i\}_{i=\overline{1,n}}$, $\hat{E}=\{\hat{\mathbf{e}}_i\}_{i=\overline{1,n}}$ sets of all $\mathbf{R},\hat{\mathbf{R}}$ orthonormalized eigenvectors

$\psi_i=\{\lambda_i,\mathbf{e}_i\}$ - i-th eigenpair of λ_i and $\mathbf{e}_i$

$\Psi_1=\bigcup_{i=1}^{n}\mathbf{\Psi}_i$, $\mathbf{\Psi}_i=\{\psi_i,\hat{\psi}_i\}, i=\overline{1,n}$ - a set of i-th eigenpairs $\psi_i,\hat{\psi}_i$ of $\mathbf{R},\hat{\mathbf{R}}$

$\Psi_2=\bigcup_{i,j=1}^{n}\mathbf{\Psi}_{ij}$, $\mathbf{\Psi}_{ij}=\{\mathbf{\Psi}_i,\mathbf{\Psi}_j\}=\{\{\psi_i,\hat{\psi}_i\},\{\psi_j,\hat{\psi}_j\}\}$ with $i,j=\overline{1,n}$ - a set of two

pairs of eigenpairs $\mathbf{\Psi}_i,\mathbf{\Psi}_j$ of $\mathbf{R},\hat{\mathbf{R}}$

$\Psi_n=\mathbf{\Psi}_{\overline{1,n}}=\{\{\psi_1,\hat{\psi}_1\}\ldots\{\psi_n,\hat{\psi}_n\}\}$ be a set of n eigenpairs $\mathbf{\Psi}_1,\mathbf{\Psi}_2,\ldots\mathbf{\Psi}_n$

$\det \mathbf{R}$ - determinant of matrix $\mathbf{R}$

$\text{cond}\,\mathbf{R}$ - condition of matrix $\mathbf{R}$

Θ - Theta Criteria

$\mathbf{T}$ - limited linear, self-conjugated integral matrix from space $L_2(\mathbf{X},\mu)$ into $L_2(\mathbf{X},\mu)$

$\mathbf{K}(.,.) \in L_2(\mathbf{X}\times\mathbf{X},\ \mu\times\mu)$ - matrix's kernel

$\|\ \|_2$ - Euclidean norm

$LINK$ - linkage coefficient between matrices blocks

$\boldsymbol{R} = \{\mathbf{R}_i\}_{i=0}^{N},\ \text{rank}(\mathbf{R}) = n, \text{rank}(\hat{\mathbf{R}}) = n, \det(\mathbf{R}_i) \geq 0$ - the sequence of symmetrical positively defined matrices

$\Delta(\det(\mathbf{R}))$ - forward difference of the determinants of $\mathbf{R}, \hat{\mathbf{R}}$

$\Delta(\text{cond}(\mathbf{R}))$ - forward difference of the condition numbers of $\mathbf{R}, \hat{\mathbf{R}}$

Table of Contents

CHAPTER II. Theta Criteria Formal Study

CHAPTER III. Theta Criteria Numerical Study

CHAPTER IV. Applications

CHAPTER I

Existing Methods for Multivariate Data Processing

There are three major mathematical and statistical software packages to process multivariate data:

- MATLAB® [1]
- SAS® [2]
- SPSS® [3]

These software packages are based on **Multivariate General Linear Hypothesis** (MGLH) [4]:

- All dataset variables are linear
- Additive
- Relationships models are linear series of weighted terms.

The MGLH is implemented using the following procedures:

- Multiple Regression
- Discriminant Function Analysis
- Canonical Analysis
- Principle Components Analysis
- Formal linear algebra methods

We will now discuss these procedures in detail.

Multiple Regression Equation

$y = b_1x_1 + b_2x_2 + \dots + b_nx_n + c$

In this equation, y is a dependent variable, b_i - regression coefficients and x_i - independent variables. This equation evaluates y variance proportion at a significant level and x_i relative predictive importance. This method evaluates dependent variable based on independent variable values.

Discriminant Function Analysis

This method determines which variables discriminate between two or more groups on covariance matrix of group variances and co-variances. Then one of the test statistics for eigenvalue analysis, such as Wilks' Lambda, is used. This method is identical to multivariate analysis of variance or MANOVA. For several groups, additional Discriminant functions can be used.

Canonical Analysis

This method uses optimal variables combination for multiple group Discriminant analysis. The first function is the most informative description, the second is second most, and so on. The functions ought to be independent or orthogonal. The canonical correlation analysis is based primarily on canonical roots or eigenvalues.

Factor Structure Method

This method analyzes correlations of variables and interpretes the Discriminant functions' values. This method places heavy emphasis on results interpretation and will not be reviewed here.

Principle Components Analysis (PCA)

This method has been used to estimate the dataset variance in terms of principle components. The method goals are to reduce data dimensionality, define the most informative components and noise filtering. The standard normalization procedure removes noise, stabilizes the data. Regrettably, this method has limited efficiency as data structure identification tool. The PCA defines mutually-orthogonal or uncorrelated projections set. For square and symmetric matrix with ordered eigenvalues, the first principal component direction coincides with 1st eigenvector direction. The second principal component direction coincides with direction of 2nd eigenvector direction. The procedure iterates until satisfactory accuracy has been achieved.

For symmetric matrix, the eigenvalue and eigenvectors can be found by a Householder reduction procedure and QL algorithm. For non-square or non-symmetric data matrix A, the singular value decomposition $U\ V'$ of A can be formed. Here matrix V contains the eigenvectors, and the squared diagonal matrix U contains the eigenvalues [5], [6].

Formal Linear Algebra Methods

These methods use various norms, determinant, trace and condition to evaluate the matrices distance. Nearly all of those criteria can be represented as various functions of eigenvalues [7], [8].

Theta Criteria

According to Spectral and Hilbert Theorems, the whole sets of eigenvalues & eigenvectors or eigenvalues & eigenfunctions fully describe matrix or operator. Our methods (Theta Criteria) are constructed from whole sets of eigenvalues & eigenvectors or eigenvalues & eigenfunctions. In this scenario, Theta Criteria methods are more optimal for multivariate applications than existing methods. We studied the Theta Criteria in detail and found these methods to be more precise and accurate than existing methods [9], [10].

Let us assume that Spectral Theorem conditions are fulfilled and symmetrical operator / matrix $\mathbf{R}$ can be diagonalized. Also, orthonormalized basis of $\mathbf{R}$ exists consisting of its eigenvectors. In addition, each eigenvalue of $\mathbf{R}$ is real.

Let $\mathbf{R}$ and $\hat{\mathbf{R}}$ be symmetrical matrices or operators. Let us construct set of criteria $\Theta = \{\Theta_i(\mathbf{R},\hat{\mathbf{R}}), \Theta_{ij}(\mathbf{R},\hat{\mathbf{R}}), \Theta_{i...k}(\mathbf{R},\hat{\mathbf{R}})\}$ $i, j, k = \overline{1,n}$, which can converge on $L_2(X,\mu)$, $L_2(X \times X, \mu \times \mu)$. Such criteria will reflect the geometrical changes on some the elements of Ψ_1, Ψ_2 ... or Ψ_n.

Let evaluate 1st differences $\varphi_i = \lambda_i \mathbf{e}_i - \hat{\lambda}_i \hat{\mathbf{e}}_i$ between weighted eigenvectors $\lambda_i \mathbf{e}_i$ and $\hat{\lambda}_i \hat{\mathbf{e}}_i$. Their Euclidean norm, or Θ_i criteria $\Theta_i = \|\varphi_i\|_2$ can serve as closeness criteria between eigenpairs $\{\lambda_i, \mathbf{e}_i\}$ and $\{\hat{\lambda}_i, \hat{\mathbf{e}}_i\}$ (Figure 1.1.) Analogously, $\Theta_{ij} = \|\varphi_i\|_2 + \|\varphi_j\|_2$ can be Θ_{ij} criteria between pairs of eigenpairs $\{\lambda_i, \mathbf{e}_i\}$, $\{\hat{\lambda}_i, \hat{\mathbf{e}}_i\}$ and $\{\lambda_j, \mathbf{e}_j\}$, $\{\hat{\lambda}_j, \hat{\mathbf{e}}_j\}$. At last, $\Theta_{\overline{1,n}} = \sum_{i=1}^{n} \|\varphi_i\|$ be $\Theta_{\overline{1,n}}$ criteria on all eigenpairs $\{\lambda_i, \mathbf{e}_i\}_{i=1}^{n}$.

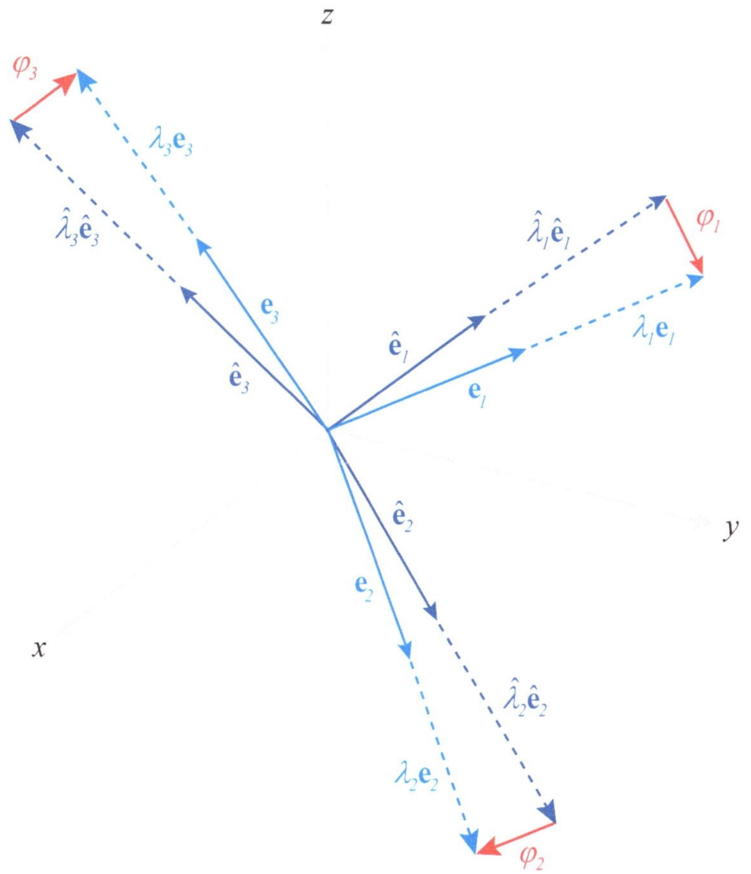

Figure 1.1.

Theta Criteria Properties

We have found that Theta Criteria are norms. These methods are positive, homogeneous, positively defined and satisfy triangle inequality. The Theta Criteria can be transformed to matrix norm and trace differences.

We formulated distinction types hypotheses for positively defined matrices $\mathbf{R}$ and $\hat{\mathbf{R}}$. Then we evaluated accuracy of Theta Criteria and $|\det \mathbf{R} - \det \hat{\mathbf{R}}|$ or $|\operatorname{cond} \mathbf{R} - \operatorname{cond} \hat{\mathbf{R}}|$ for very close matrices and for ill-defined matrices. Several Theta Criteria were significantly more accurate than $|\det \mathbf{R} - \det \hat{\mathbf{R}}|$ or $|\operatorname{cond} \mathbf{R} - \operatorname{cond} \hat{\mathbf{R}}|$. Further research is required to obtain functional representation between distinction hypotheses types and Theta Criteria optimal type(s).

Summary

Existing application for multivariate data set processing, such as MATLAB® [1], SAS® [2] and SPSS® [3] utilize Multiple Regression Procedure, Discriminant Function Analysis, Canonical Analysis and Principle Components Analysis. Those methods are appropriate for initial stage of data analysis when distinction hypotheses about specific application are not formulated or not adequately described. If distinction hypotheses were established, then formal linear algebra methods or Theta Criteria can be applied for in-depth application analysis.

The formal linear algebra methods are straightforward by utilizing only matrices' eigenvalues. If the application accuracy specifications are moderate, then these methods will be sufficient. Regrettably, formal linear algebra methods have limited accuracy for complex or ill-defined applications.

If, however, the multivariate application is ill-defined or requires high accuracy, then Theta Criteria deserve serious consideration.

References:

1. MATLAB® USER GUIDE

   www.mathworks.com/access/helpdesk/help/techdoc/data_analysis/data_analysis.html

2. SAS® / STAT® User Guide

   www.id.unizh.ch/software/unix/statmath/sas/sasdoc/stat/

3. SPSS® / SCS® Documentation Guide

   www.spss.com/spss/data_analysis.htm

4. L. Lebart, A. Morineau and K. Warwick (1984), Multivariate Descriptive Statistical Analysis, New York: John Wiley & Sons, Inc.
5. G. Golub, C. Van Loan (1996) Matrix Computations: The John Hopkins University Press.
6. G. Stewart (1998), Matrix Decompositions, Philadelphia: SIAM.
7. S. Lang (1993), Real and Functional Analysis, New York, Berlin: Springer-Verlag
8. K. Mardia, J. Kent and J. Bibby (1979), *Multivariate Analysis,* London: Academic Press.
9. V. Vasil'ev, V. Halitsky (same person as S. Halitsky) and V. Revenko. Estimation of Correlations between Groups of Parameters of a Multidimensional Plant. Soviet Automat. Control. 1975 No. 2, 9-15.

CHAPTER II

Theta Criteria Formal Study

The Theta Criteria methods for positively defined $n \times n$ matrices were introduced in [1] - [4]. Those criteria have been constructed on norms of differences of matrices ordered weighted eigenvectors. We will now study Theta Criteria properties in depth.

2.1. Formal Definitions

Let R is real numbers field, R^n-finite linear vector space over R,

$\mathbf{x} \in R^n, \Leftrightarrow \mathbf{x} = \begin{bmatrix} x_1 \\ \vdots \\ x_n \end{bmatrix}, x_i \in R$ and $R^{n \times n}$-set of all positively defined matrices

of order n. Let block matrices $\mathbf{R}, \hat{\mathbf{R}} \in R^{n \times n}$; $\operatorname{rank}(\mathbf{R}) = \operatorname{rank}(\hat{\mathbf{R}}) = n$,

$$\mathbf{R} = \begin{bmatrix} \mathbf{A}_1 & \mathbf{B}^T \\ \mathbf{B} & \mathbf{A}_2 \end{bmatrix}, \hat{\mathbf{R}} = \begin{bmatrix} \hat{\mathbf{A}}_1 & \hat{\mathbf{B}}^T \\ \hat{\mathbf{B}} & \hat{\mathbf{A}}_2 \end{bmatrix} \tag{1}$$

sub matrices $\mathbf{A}_1 = [a_{ij}]_{i,j=\overline{1,m}}$, $\mathbf{A}_2 = [a_{ij}]_{i,j=\overline{m+1,\, n}}$, $\mathbf{B} = [a_{ij}]_{i=\overline{m+1,n};j=\overline{1,m}}$ and

$\hat{\mathbf{A}}_1 = [\hat{a}_{ij}]_{i,j=\overline{1,m}}$, $\hat{\mathbf{A}}_2 = [\hat{a}_{ij}]_{i,j=\overline{m+1,n}}$, $\hat{\mathbf{B}} = [\hat{a}_{ij}]_{i=\overline{m+1,n};j=\overline{1,m}}$.

Let $\Lambda = \{\lambda_i\}_{i=\overline{1,n}}$, $\hat{\Lambda} = \{\hat{\lambda}_i\}_{i=\overline{1,n}}$ - sets of all ordered eigenvalues of $\mathbf{R}, \hat{\mathbf{R}}$:

$$\lambda_i > \lambda_j \;,\quad \hat{\lambda}_i > \hat{\lambda}_j \;,\quad i < j \;,\quad \forall\; i,j = \overline{1,n}, \tag{2}$$

and $E = \{\mathbf{e}_i\}_{i=\overline{1,n}}$ and $\hat{E} = \{\hat{\mathbf{e}}_i\}_{i=\overline{1,n}}$ sets of all $\mathbf{R}, \hat{\mathbf{R}}$ orthonormalized eigenvectors.

Let $\boldsymbol{\psi}_i = \{\lambda_i, \mathbf{e}_i\}$ - eigenpair of λ_i and $\mathbf{e}_i$ and $\Psi_1 = \bigcup_{i=1}^{n} \boldsymbol{\Psi}_i$, with

$\boldsymbol{\Psi}_i = \{\boldsymbol{\psi}_i, \hat{\boldsymbol{\psi}}_i\}, i = \overline{1,n}\;\; i = \overline{1,n}$ - a set of pairs of eigenpairs of i-th eigenvalues and eigenvectors of $\mathbf{R}, \hat{\mathbf{R}}$.

Let $\Psi_2 = \bigcup_{i,j=1}^{n} \boldsymbol{\Psi}_{ij}$, $\boldsymbol{\Psi}_{ij} = \{\boldsymbol{\Psi}_i, \boldsymbol{\Psi}_j\} = \{\{\boldsymbol{\psi}_i, \hat{\boldsymbol{\psi}}_i\}, \{\boldsymbol{\psi}_j, \hat{\boldsymbol{\psi}}_j\}\}$ with $i, j = \overline{1,n}$ be a set of two pairs of eigenpairs of i-th and j-th eigenvalues and eigenvectors of $\mathbf{R}, \hat{\mathbf{R}}$ and $\Psi_n = \boldsymbol{\Psi}_{\overline{1,n}} = \{\{\boldsymbol{\psi}_1, \hat{\boldsymbol{\psi}}_1\} \ldots \{\boldsymbol{\psi}_n, \hat{\boldsymbol{\psi}}_n\}\}$ has been composed on n eigenpairs $\boldsymbol{\Psi}_1, \boldsymbol{\Psi}_2, \ldots \boldsymbol{\Psi}_n$.

2.2. Known Matrices Closeness Criteria

The forward differences Δ of the determinants and condition numbers were used as matrices closeness criteria [5] - [9]:

$$\Delta(\det(\mathbf{R}) = \left|\det(\mathbf{R}) - \det(\hat{\mathbf{R}})\right| = \left|\prod_{i=1}^{n} \lambda_i - \prod_{i=1}^{n} \hat{\lambda}_i\right| \tag{3}$$

$$\Delta(\mathrm{cond}(\mathbf{R})) = \left|\mathrm{cond}(\mathbf{R}) - \mathrm{cond}(\hat{\mathbf{R}})\right| = \left|\lambda_1/\lambda_n - \hat{\lambda}_1/\hat{\lambda}_n\right| \tag{4}$$

The Θ criteria of $\mathbf{R}, \hat{\mathbf{R}}$ has been introduced in [4]:

$$\Theta\left(\mathbf{R}, \hat{\mathbf{R}}\right) = \sum_{i=1}^{n} \left\| \lambda_i \mathbf{e}_i - \hat{\lambda}_i \hat{\mathbf{e}}_i \right\|. \qquad (5)$$

2.3. The Hilbert Theorem

Let $\mathbf{T}$ be a limited linear, self-conjugated integral matrix from space $L_2(\mathbf{X}, \mu)$ into $L_2(\mathbf{X}, \mu)$ and

$(\mathbf{T}f)(x) = \int_x \mathbf{K}(x, y) f(y) \mu(Dy)$, $\mathbf{K}(x, y) = \mathbf{K}'(y, x)$; $f \in L_2(\mathbf{X}, \mu)$ where

$\mathbf{K}(.,.) \in L_2(\mathbf{X} \times \mathbf{X}, \ \mu \times \mu)$ is the matrix's kernel. There exist $\mathbf{T}, \mathbf{K}$ representations on orthonormalized matrices of eigenfunctions $\{\varphi_i(x)\}$ and eigenvalues $\{\lambda_i(x)\}$ of $\mathbf{T}$ as follows:

$$(\mathbf{T}f)(x) = \sum_i \lambda_i (f, \varphi_i) \varphi_i, \qquad f \in L_2(\mathbf{X}, \mu) \qquad (6)$$

$$\mathbf{K}(x, y) = \sum_i \lambda_i \varphi_i(x) \varphi_i', \qquad \lambda_i \neq 0 \qquad (7)$$

The series are converging on norms $L_2(X, \mu)$, $L_2(X \times X, \mu \times \mu)$ respectively [10] - [12].

2.4. Theta Criteria or Θ Criteria Construction

Let construct Θ criteria between $\mathbf{R}, \hat{\mathbf{R}}$, or $\Theta(\mathbf{R}, \hat{\mathbf{R}})$, which can converge on $L_2(X, \mu)$, $L_2(X \times X, \mu \times \mu)$. Such criteria will reflect the geometrical changes on some the elements of Ψ_1, Ψ_2 ... or Ψ_n. The proper choice of Θ criteria depends on a priori information about $\mathbf{R},\hat{\mathbf{R}}$ structures and their distinction type. If all elements of $\mathbf{\Psi}$ have changed, then $\Theta_{\overline{1,n}}(\mathbf{R},\hat{\mathbf{R}})$ is appropriate choice. If only $\{\psi_i,\hat{\psi}_i\}$ and $\{\psi_j,\hat{\psi}_j\}$ have changed, then $\Theta_{ij}(\mathbf{R},\hat{\mathbf{R}})$ is acceptable. Now we can formulate several hypotheses about matrices $\mathbf{R},\hat{\mathbf{R}}$ differences.

Hypothesis I: The matrices $\mathbf{R},\hat{\mathbf{R}}$ distinctions can be represented by geometrical differences between ψ_i and $\hat{\psi}_i$ of $\mathbf{\Psi}_i=\{\psi_i,\hat{\psi}_i\}$.

Then Euclidean norm$\|\ \|_2$ of $\varphi_i=\lambda_i\mathbf{e}_i-\hat{\lambda}_i\hat{\mathbf{e}}_i$ can serve as $\Theta(\mathbf{A},\hat{\mathbf{A}})$ or

$$\Theta_i=\|\varphi_i\|_2=\|\lambda_i\mathbf{e}_i-\hat{\lambda}_i\hat{\mathbf{e}}_i\|_2. \qquad (8)$$

Hypothesis II: The matrices $\mathbf{R},\hat{\mathbf{R}}$ distinction is represented by geometrical differences between $\{\psi_i,\hat{\psi}_i\}$ and $\{\psi_j,\hat{\psi}_j\}$ of $\mathbf{\Psi}_{ij}=\{\{\psi_i,\hat{\psi}_i\},\{\psi_j,\hat{\psi}_j\}\}$.

Then the sum of $\|\varphi_i\|_2$ and $\|\varphi_j\|_2$ can serve as $\Theta(\mathbf{A},\hat{\mathbf{A}})$:

$$\Theta(\mathbf{R},\hat{\mathbf{R}})=\Theta_{ij}(\mathbf{R},\hat{\mathbf{R}})=\|\varphi_i\|_2+\|\varphi_j\|_2=\|\lambda_i\mathbf{e}_i-\hat{\lambda}_i\hat{\mathbf{e}}_i\|_2+\|\lambda_j\mathbf{e}_j-\hat{\lambda}_j\hat{\mathbf{e}}_j\|_2 \qquad (9)$$

Hypothesis III: The matrices $\mathbf{R},\hat{\mathbf{R}}$ distinction is represented by geometrical differences between $\{\psi_1,\hat{\psi}_1\},\{\psi_2,\hat{\psi}_2\}...\{\psi_n,\hat{\psi}_n\}$ of $\mathbf{\Psi}_{\overline{1,n}}=\{\{\psi_1,\hat{\psi}_1\},\{\psi_2,\hat{\psi}_2\},...\{\psi_n,\hat{\psi}_n\}\}$.

Then the sum of $\|\varphi_1\|_2$, $\|\varphi_2\|_2 \ldots \|\varphi_n\|_2$ can serve as $\Theta(\mathbf{R},\hat{\mathbf{R}})$:

$$\Theta_{\overline{1,n}} = \sum_{i=1}^{n} \|\varphi_i\|_2 . \tag{10}$$

According to [9], a real-valued function $\|x\|$ on linear space X, $x \in \mathrm{X}$ is *norm on* X, if

$\|x\| \geq 0$ (Positivity) (11)

$\|x + y\| \geq \|x\| + \|y\|$ (Triangle inequality) (12)

$\|\alpha x\| = |\alpha|\|x\|$ (Homogeneity) (13)

$\|x\| = 0$ if and only if $x = 0$. (Positive definiteness) (14)

2.5. Theta Criteria Properties

Theorem 1. (Positivity).

The criteria $\Theta_{\overline{ik}}(\mathbf{R},\hat{\mathbf{R}}) \geq 0$.

Proof: From Θ criteria definition and Euclidean norm properties

$$\Theta_{\overline{jk}} = \sum_{i=j}^{k} \left\|\lambda_i \mathbf{e}_i - \hat{\lambda}_i \hat{\mathbf{e}}_i\right\|_2 = \sum_{i=j}^{k} \|\varphi_i\|_2 \geq 0.$$

Q.E.D.

Theorem 2. (Triangle inequality)

If $\theta_1 = \Theta_{\overline{j,k}}(\mathbf{R},\hat{\mathbf{R}})$, $\theta_2 = \Theta_{\overline{j,k}}(\mathbf{R},\widetilde{\mathbf{R}})$, $\theta_3 = \Theta_{\overline{j,k}}(\hat{\mathbf{R}},\widetilde{\mathbf{R}})$, $\mathbf{R},\hat{\mathbf{R}},\widetilde{\mathbf{R}} \in R^{n\times n}$, then

$\theta_1 + \theta_2 \geq \theta_3$.

Proof: According to $\Theta(\mathbf{A},\hat{\mathbf{A}})$ definition,

$$\theta_1 = \sum_{i=j}^{k} \left\| \lambda_i \mathbf{e}_i - \hat{\lambda}_i \hat{\mathbf{e}}_i \right\|_2, \theta_2 = \sum_{i=j}^{k} \left\| \lambda_i \mathbf{e}_i - \widetilde{\lambda}_i \widetilde{\mathbf{e}}_i \right\|_2, \theta_3 = \sum_{i=j}^{k} \left\| \hat{\lambda}_i \hat{\mathbf{e}}_i - \widetilde{\lambda}_i \widetilde{\mathbf{e}}_i \right\|_2.$$

Since $\mathbf{e}_i, \hat{\mathbf{e}}_i, \widetilde{\mathbf{e}}_i \in R^n$ for $\forall i = \overline{1,n}$, the vectors $\lambda_i \mathbf{e}_i, \hat{\lambda}_i \hat{\mathbf{e}}_i, \widetilde{\lambda}_i \widetilde{\mathbf{e}}_i \in R^n$ for $\forall i = \overline{1,n}$.

Then $\sum_{i=j}^{k} \left\| \lambda_i \mathbf{e}_i - \hat{\lambda}_i \hat{\mathbf{e}}_i \right\|_2 + \sum_{i=j}^{k} \left\| \lambda_i \mathbf{e}_i - \widetilde{\lambda}_i \widetilde{\mathbf{e}}_i \right\|_2 \geq \sum_{i=j}^{k} \left\| \hat{\lambda} \hat{\mathbf{e}}_i - \widetilde{\lambda}_i \widetilde{\mathbf{e}}_i \right\|_2$.

Q.E.D.

Theorem 3. (Homogeneity):

$\Theta\left(\alpha\mathbf{R},\alpha\hat{\mathbf{R}}\right) = |\alpha| \Theta\left(\mathbf{R},\hat{\mathbf{R}}\right)$, where $\alpha \in R$.

Proof: Since $\mathbf{e}_i, \hat{\mathbf{e}}_i \in R^n$ and $\lambda_i, \hat{\lambda}_i \in R$ for $\forall i = \overline{1,n}$,

$$\Theta_{\overline{jk}}\left(\alpha\mathbf{R},\alpha\hat{\mathbf{R}}\right) = \sum_{i=j}^{k} \left\| \alpha\lambda_i \mathbf{e}_i - \alpha\hat{\lambda}_i \hat{\mathbf{e}}_i \right\|_2 = \sum_{i=j}^{k} |\alpha| \left\| \lambda_i \mathbf{e}_i - \hat{\lambda}_i \hat{\mathbf{e}}_i \right\|_2$$

Q.E.D.

Theorem 4. (Positive definiteness)

$\Theta_{\overline{jk}}(\mathbf{R},\hat{\mathbf{R}}) = 0$ if and only if $\mathbf{R} = \hat{\mathbf{R}}$.

Proof: Let $\mathbf{R} = \hat{\mathbf{R}}$. The Θ_{jk} criteria is $\Theta_{\overline{jk}} = \sum_{i=j}^{k} \left\| \lambda_i \mathbf{e}_i - \hat{\lambda}_i \hat{\mathbf{e}}_i \right\|_2$. The i-th component of Θ_{jk} is

$\Theta_{jk}{}'^{\,i} = \lambda_i \mathbf{e}_i - \hat{\lambda}_i \hat{\mathbf{e}}_i$. According to [2], [3] and Hilbert Theorem

$\lambda_i = \hat{\lambda}_i, \;\; \mathbf{e}_i = \hat{\mathbf{e}}_i, \;\; \forall i = \overline{1,n}$ and $\Theta_{jk}{}'^{\,i} = 0$.

Then $\Theta_{\overline{jk}}(\mathbf{R},\hat{\mathbf{R}}) = 0$ is true, because index i is arbitrary.

Let $\Theta_{\overline{j,k}}(\mathbf{R},\hat{\mathbf{R}}) = \sum_{i=j}^{k} \left\| \lambda_i \mathbf{e}_i - \hat{\lambda}_i \hat{\mathbf{e}}_i \right\|_2 = 0$.

Then we will receive the system of $k - j$ equations

$$\begin{cases} \left\| \lambda_j \mathbf{e}_j - \hat{\lambda}_j \hat{\mathbf{e}}_j \right\|_2 = 0 \\ \ldots\ldots \\ \left\| \lambda_k \mathbf{e}_k - \hat{\lambda}_k \hat{\mathbf{e}}_k \right\|_2 = 0 \end{cases}$$

with solution $\lambda_i = \hat{\lambda}_i, \;\; \mathbf{e}_i = \hat{\mathbf{e}}_i, \;\; \forall i = \overline{j,k}$. According to the Hilbert theorem, for each $\mathbf{R}$ and $\hat{\mathbf{R}}$ exist unique $\{\Lambda, E\}$ and $\{\hat{\Lambda}, \hat{E}\}$. If $\Theta_{\overline{jk}}(\mathbf{R},\hat{\mathbf{R}}) = 0$, then $\Lambda \equiv \hat{\Lambda}$, $E \equiv \hat{E}$ and $\mathbf{R} = \hat{\mathbf{R}}$.

Conclusion: The criteria $\Theta(\mathbf{R},\hat{\mathbf{R}})$ is a norm on $R^{n \times n}$.

Theorem 5. (symmetry)

If $\theta = \Theta_{\overline{j,k}}(\mathbf{R},\hat{\mathbf{R}})$ and $\theta' = \Theta_{\overline{j,k}}(\hat{\mathbf{R}},\mathbf{R})$ then $\theta = \theta'$.

Proof: If $\mathbf{R}$ and $\hat{\mathbf{R}}$ switch places in $\theta, \hat{\theta}$, then $\theta = \theta'$.

Theorem 6.

If $\mathbf{e}_1 = \hat{\mathbf{e}}_1$, $\psi_i \equiv \hat{\psi}_i, i = \overline{2,n}$, then the $\Theta(\mathbf{R},\hat{\mathbf{R}})$ is the matrix norm difference

$$\Theta\left(\mathbf{R},\hat{\mathbf{R}}\right) = \left|\lambda_1 - \hat{\lambda}_1\right|.$$

Proof:

Criteria $\Theta(\mathbf{R},\hat{\mathbf{R}}) = \sum_{i=1}^{n} \left\|\lambda_i \mathbf{e}_i - \hat{\lambda}_i \hat{\mathbf{e}}_i\right\|_2 = \left\|\lambda_1 \mathbf{e}_1 - \hat{\lambda}_1 \hat{\mathbf{e}}_1\right\|_2 \Big|_{\psi_i \equiv \hat{\psi}_i, i=\overline{2,n}} = \left|\lambda_1 - \hat{\lambda}_1\right|_{\mathbf{e}_1 = \hat{\mathbf{e}}_1}$.

Theorem 7.

If $\mathbf{e}_i \equiv \hat{\mathbf{e}}_i, i = \overline{1,n}$, $\lambda_i \geq \hat{\lambda}_i$, $\forall i = \overline{1,n}$ then $\Theta\left(\mathbf{R},\hat{\mathbf{R}}\right) = \mathrm{tr}(\mathbf{R}) - \mathrm{tr}(\hat{\mathbf{R}})$.

Proof:

From $\mathbf{e}_i \equiv \hat{\mathbf{e}}_i, \|\mathbf{e}_i\| = \|\hat{\mathbf{e}}_i\| = 1$, $i = \overline{1,n}$, and $\lambda_i \geq \hat{\lambda}_i$, $i = \overline{1,n}$ we received:

$$\Theta(\mathbf{R},\hat{\mathbf{R}}) = \sum_{i=1}^{n} \left\|\lambda_i \mathbf{e}_i - \hat{\lambda}_i \hat{\mathbf{e}}_i\right\|_2 = \sum_{i=1}^{n} \left|\lambda_i - \hat{\lambda}_i\right|_{\mathbf{e}_i \equiv \hat{\mathbf{e}}_i, \|\mathbf{e}_i\| = \|\hat{\mathbf{e}}_i\| = 1, i=\overline{1,n}} = \mathrm{tr}(\mathbf{R}) - \mathrm{tr}(\hat{\mathbf{R}})\Big|_{\lambda_i \geq \hat{\lambda}_i, i=\overline{1,n}}$$

References:

1. V. Vasil'ev, V. Halitsky (same person as S. Halitsky) and V. Revenko. Estimation of Correlations between Groups of Parameters of a Multidimensional Plant. Soviet Automat. Control. 1973, no. 2, 9-15.
2. V. Halitsky (same person as S. Halitsky). Shortening of description of a dynamic control system with given constraints. The Proceedings of Ukrainian Academy of Science, Type A, 1985, #9, pp. 65-67.
3. S. Halitsky. About closeness criteria of positively-defined operators. The Proceedings of 1st Bernoulli Congress on Mathematical Statistics, Vol. 2, pp. 926-927, "Nauka," Moscow, 1986.
4. S. Halitsky. The closeness criteria for positively-defined operators Express-Printing Edition, #97-4, pp. 1-17, 1997, Russian National Academy of Science.
5. H. Weyl. The Classical Groups, Their Invariants and Representations. Princeton, 1999.
6. G. W. Stewart. Matrix algorithms. Vol. 1. Basic Decompositions. SIAM, 1998.

7. G. Golub, C. F. Van Loan. Matrix Computations. The John Hopkins Univ. Press, 3rd Ed., 1996.
8. N. J. Higham. Accuracy and Stability of Numerical Algorithms. SIAM, 1996.
9. V.B. Korotkov, Hilbert-Schmidt integral operator. Mathematics Encyclopedia, Vol. 1, p. 975. Soviet Encyclopedia, Moscow, 1977.
10. B.M. Bredichin, Hilbert-Schmidt integral operator. Encyclopaedia of Mathematics, Springer 2002.
11. N. Akhieser, I. Glazman. Theory of Linear Operators in Hilbert Space. Dover Ed., 1003.
12. R. Horn, C. Johnson. Matrix Analysis. Cambridge University Press. 1985.

CHAPTER III

Theta Criteria Numerical Study

3.1. Numerical Experiments Details

Let $\mathbf{R} = \begin{bmatrix} \mathbf{A}_1 & \mathbf{B}^T \\ \mathbf{B} & \mathbf{A}_2 \end{bmatrix}$, $\hat{\mathbf{R}} = \begin{bmatrix} \mathbf{A}_1 & \mathbf{0} \\ \mathbf{0} & \mathbf{A}_2 \end{bmatrix}$ be two positively defined matrices.

Let construct on $\mathbf{R}$ the sequence $\boldsymbol{R}$ of symmetrical positively defined matrices: $\boldsymbol{R} = \{\mathbf{R}_i\}_{i=0}^{N}$, $\mathrm{rank}(\mathbf{R}) = n$, $\mathrm{rank}(\hat{\mathbf{R}}) = n$, $\det(\mathbf{R}_i) \geq 0$, with

$\mathbf{R}_0 = \mathbf{R}$, $\mathbf{R}_i = \begin{bmatrix} \mathbf{A}_1 & \mathbf{B}_i^T \\ \mathbf{B}_i & \mathbf{A}_2 \end{bmatrix}$

$\mathbf{B}_i = LINK^i * \mathbf{B}$, $i = \overline{1, N}$; $0 < LINK < 1$; $\lim_{i\to\infty} \mathbf{B}_i = \mathbf{0}$; $\lim_{i\to\infty} \Theta(\mathbf{R}_i, \hat{\mathbf{R}}) = 0$.

Numerical results have been obtained on MATLAB Version 6.5 for matrices $\mathbf{R}_i, rank(\mathbf{R}_i) = 4,5,10$. Accuracies of algorithms have been verified by methods from [6] - [8]. The matrix block linkage $LINK$ and sequence $\boldsymbol{R}$ cardinality N are : $LINK_{opt} = 0.25$, $N_{opt} = 100$. We assumed that $\mathbf{R}_i$ and $\hat{\mathbf{R}}$ distinction on sequence $\boldsymbol{R}$ can be represented by criteria $\Theta = \left\{\left\{\Theta_j(\mathbf{R}_i, \hat{\mathbf{R}})\right\}_{i=1,N}\right\}_{j=1}^{n}$, $\Delta\det(\mathrm{R}) = \left\{\det(\mathbf{R}_i) - \det(\hat{\mathbf{R}})\right\}_{i=\overline{1,N}}$, $\Delta\,\mathrm{cond}(\mathrm{R}) = \left\{\mathrm{cond}(\mathbf{R}_i) - \mathrm{cond}(\hat{\mathbf{R}})\right\}_{i=\overline{1,N}}$.

The adequate $\Theta_i, i = \overline{1,n}$ criteria compared with $\Delta\det(\mathrm{R})$ and $\Delta\,\mathrm{cond}(\mathrm{R})$ criteria in logarithmic scale.

3.2. Sequence R of matrices $\mathbf{R}_i, rank(\mathbf{R}_i) = 4$

3.2.1. $rank(\mathbf{R}) = rank(\hat{\mathbf{R}}) = 4, rank(\mathbf{B}_i) = rank(\mathbf{A}_1) = rank(\mathbf{A}_2) = 2$.

The positively defined matrices $\mathbf{R}$ and $\hat{\mathbf{R}}$ are:

$$\mathbf{R} = \begin{bmatrix} 30 & 70 & 45 & 156 \\ 70 & 174 & 125 & 600 \\ 45 & 125 & 146 & 808 \\ 156 & 600 & 808 & 6681 \end{bmatrix}, \quad \hat{\mathbf{R}} = \begin{bmatrix} 30 & 70 & 0 & 0 \\ 70 & 174 & 0 & 0 \\ 0 & 0 & 146 & 808 \\ 0 & 0 & 808 & 6681 \end{bmatrix}$$

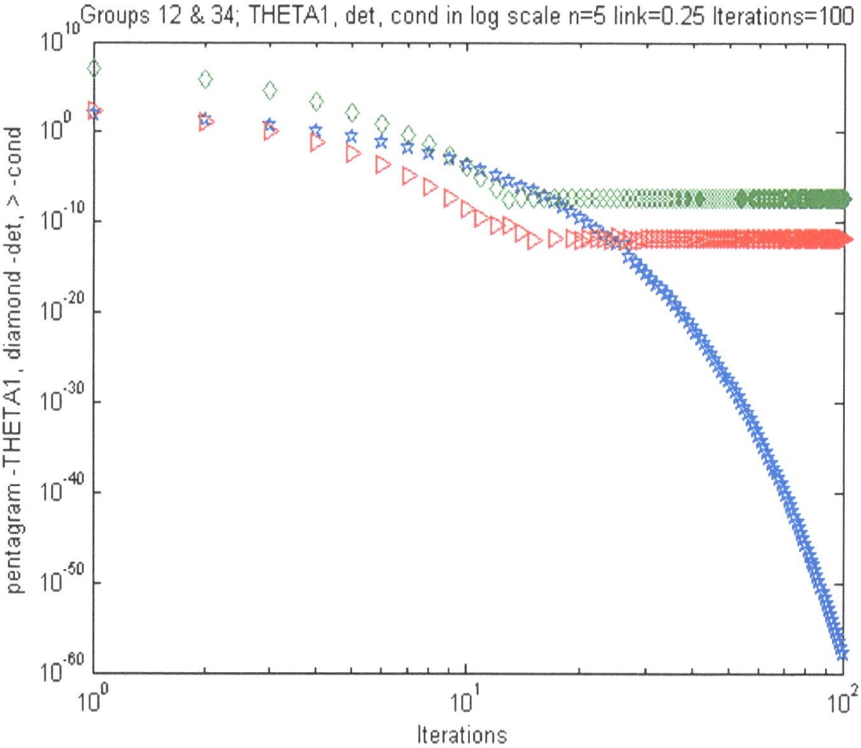

Figure 3.2.1. The results for Θ, $\Delta\,\det(\mathrm{R})$ and $\Delta\,\mathrm{cond}(\mathrm{R})$.

3.3. Sequence R of Matrices $\mathbf{R}_i, rank(\mathbf{R}_i) = 5$

3.3.1: $\text{rank}(\mathbf{R}) = \text{rank}(\hat{\mathbf{R}}) = 5, \text{rank}(\mathbf{B}_i) = \text{rank}(\mathbf{A}_1) = 2, \text{rank}(\mathbf{A}_2) = 3$.

The positively defined matrices $\mathbf{R}$ and $\hat{\mathbf{R}}$ are:

$$\mathbf{R} = \begin{bmatrix} 30 & 70 & 45 & 156 & 132 \\ 70 & 174 & 125 & 600 & 480 \\ 45 & 125 & 146 & 808 & 530 \\ 156 & 600 & 808 & 6681 & 3715 \\ 132 & 480 & 530 & 3715 & 4920 \end{bmatrix} \quad \hat{\mathbf{R}} = \begin{bmatrix} 30 & 70 & 0 & 0 & 0 \\ 70 & 174 & 0 & 0 & 0 \\ 0 & 0 & 146 & 808 & 530 \\ 0 & 0 & 808 & 6681 & 3715 \\ 0 & 0 & 530 & 3715 & 4920 \end{bmatrix}$$

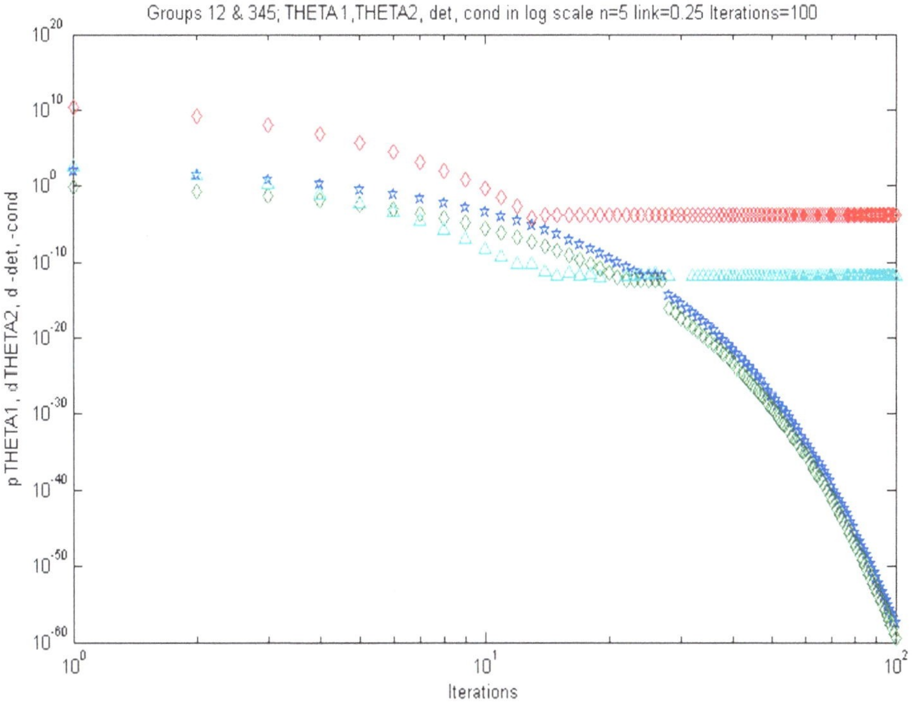

Figure 3.3.1. The results for Θ, $\Delta\,\det(\text{R})$ and $\Delta\,\text{cond}(\text{R})$.

3.3.2: $\text{rank}(\mathbf{R}) = \text{rank}(\hat{\mathbf{R}}) = 5, \text{rank}(\mathbf{B}_i) = 2, \text{rank}(\mathbf{A}_1) = 3, \text{rank}(\mathbf{A}_2) = 2$

The positively defined matrices $\mathbf{R}$ and $\hat{\mathbf{R}}$ are:

$$\mathbf{R} = \begin{bmatrix} 30 & 70 & 45 & 156 & 132 \\ 70 & 174 & 125 & 600 & 480 \\ 45 & 125 & 146 & 808 & 530 \\ 156 & 600 & 808 & 6681 & 3715 \\ 132 & 480 & 530 & 3715 & 4920 \end{bmatrix} \quad \hat{\mathbf{R}} = \begin{bmatrix} 30 & 70 & 45 & 0 & 0 \\ 70 & 174 & 125 & 0 & 0 \\ 45 & 125 & 146 & 0 & 0 \\ 0 & 0 & 0 & 6681 & 3715 \\ 0 & 0 & 0 & 3715 & 4920 \end{bmatrix}$$

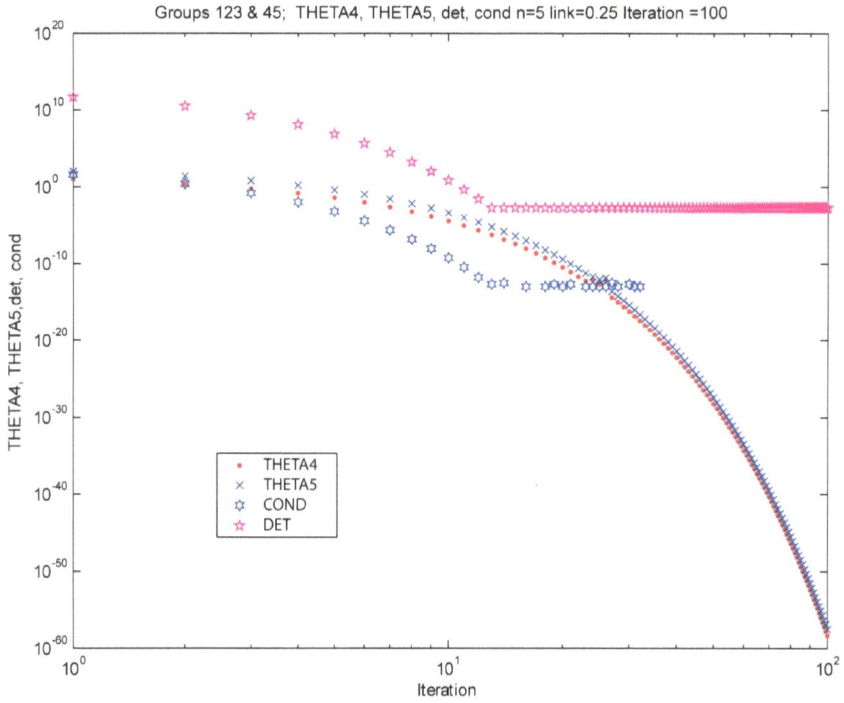

Figure 3.3.2. The results for Θ, $\Delta\,\det(\mathrm{R})$ and $\Delta\,\mathrm{cond}(\mathrm{R})$

3.4. Sequence *R* Matrices $\mathbf{R}_i, rank(\mathbf{R}_i) = 10$

The initial 10 x 10 positively defined matrix **R** is:

$$\mathbf{R} = \begin{bmatrix} 30.10 & 7.00 & 4.50 & 15.60 & 13.20 & 5.40 & 6.30 & 2.60 & 5.90 & 3.90 \\ 7.00 & 74.00 & 12.50 & 6.00 & 8.00 & 29.10 & 5.60 & 9.80 & 14.70 & 7.00 \\ 4.50 & 12.50 & 46.00 & 8.00 & 5.30 & 8.80 & 18.60 & 6.10 & 5.30 & 8.50 \\ 15.60 & 6.00 & 8.00 & 66.80 & 37.15 & 39.20 & 26.40 & 42.20 & 24.90 & 9.60 \\ 13.20 & 8.00 & 5.30 & 37.15 & 92.00 & 9.60 & 25.20 & 32.40 & 8.60 & 9.30 \\ 5.40 & 29.10 & 8.80 & 39.20 & 9.60 & 259.0 & 41.40 & 75.00 & 68.00 & 24.00 \\ 6.30 & 5.60 & 18.60 & 26.40 & 25.20 & 41.00 & 138.0 & 58.00 & 16.00 & 26.00 \\ 2.60 & 9.80 & 6.10 & 42.20 & 32.40 & 75.00 & 58.00 & 85.00 & 41.00 & 17.00 \\ 5.90 & 14.70 & 5.30 & 24.90 & 8.60 & 68.00 & 16.0 & 41.00 & 127.0 & 19.00 \\ 3.90 & 7.00 & 8.50 & 9.60 & 9.30 & 24.00 & 26.00 & 17.00 & 19.00 & 215.00 \end{bmatrix}$$

Formally, 10 x 10 matrix **R** is presented below:

$$\mathbf{R} = \begin{bmatrix} a_{11} & a_{12} & a_{13} & a_{14} & a_{15} & a_{16} & a_{17} & a_{18} & a_{19} & a_{110} \\ a_{21} & a_{22} & a_{23} & a_{24} & a_{25} & a_{26} & a_{27} & a_{28} & a_{29} & a_{210} \\ a_{31} & a_{32} & a_{33} & a_{34} & a_{35} & a_{36} & a_{37} & a_{38} & a_{39} & a_{310} \\ a_{41} & a_{42} & a_{43} & a_{44} & a_{45} & a_{46} & a_{47} & a_{48} & a_{49} & a_{410} \\ a_{51} & a_{52} & a_{53} & a_{54} & a_{55} & a_{56} & a_{57} & a_{58} & a_{59} & a_{510} \\ a_{61} & a_{62} & a_{63} & a_{64} & a_{65} & a_{66} & a_{67} & a_{68} & a_{69} & a_{610} \\ a_{71} & a_{72} & a_{73} & a_{74} & a_{75} & a_{76} & a_{77} & a_{78} & a_{79} & a_{710} \\ a_{81} & a_{82} & a_{83} & a_{84} & a_{85} & a_{86} & a_{87} & a_{88} & a_{89} & a_{810} \\ a_{91} & a_{92} & a_{93} & a_{94} & a_{95} & a_{96} & a_{97} & a_{98} & a_{99} & a_{910} \\ a_{101} & a_{102} & a_{103} & a_{104} & a_{105} & a_{106} & a_{107} & a_{108} & a_{109} & a_{1010} \end{bmatrix}$$

3.4.1: $rank(\mathbf{R}) = rank(\hat{\mathbf{R}}) = 10, rank(\mathbf{B}) = 1, rank(\mathbf{A}_1) = 1, rank(\mathbf{A}_2) = 9$.

Let $\hat{\mathbf{R}}$ will be 10-dimensional correlation matrix:

$$\hat{\mathbf{R}} = \begin{bmatrix} a_{11} & 0 & 0 & 0 & 0 & 0 & 0 & 0 & 0 & 0 \\ 0 & a_{22} & a_{23} & a_{24} & a_{25} & a_{26} & a_{27} & a_{28} & a_{29} & a_{210} \\ 0 & a_{32} & a_{33} & a_{34} & a_{35} & a_{36} & a_{37} & a_{38} & a_{39} & a_{310} \\ 0 & a_{42} & a_{43} & a_{44} & a_{45} & a_{46} & a_{47} & a_{48} & a_{49} & a_{410} \\ 0 & a_{52} & a_{53} & a_{54} & a_{55} & a_{56} & a_{57} & a_{58} & a_{59} & a_{510} \\ 0 & a_{62} & a_{63} & a_{64} & a_{65} & a_{66} & a_{67} & a_{68} & a_{69} & a_{610} \\ 0 & a_{72} & a_{73} & a_{74} & a_{75} & a_{76} & a_{77} & a_{78} & a_{79} & a_{710} \\ 0 & a_{82} & a_{83} & a_{84} & a_{85} & a_{86} & a_{87} & a_{88} & a_{89} & a_{810} \\ 0 & a_{92} & a_{93} & a_{94} & a_{95} & a_{96} & a_{97} & a_{98} & a_{99} & a_{910} \\ 0 & a_{102} & a_{103} & a_{104} & a_{105} & a_{106} & a_{107} & a_{108} & a_{109} & a_{1010} \end{bmatrix}$$

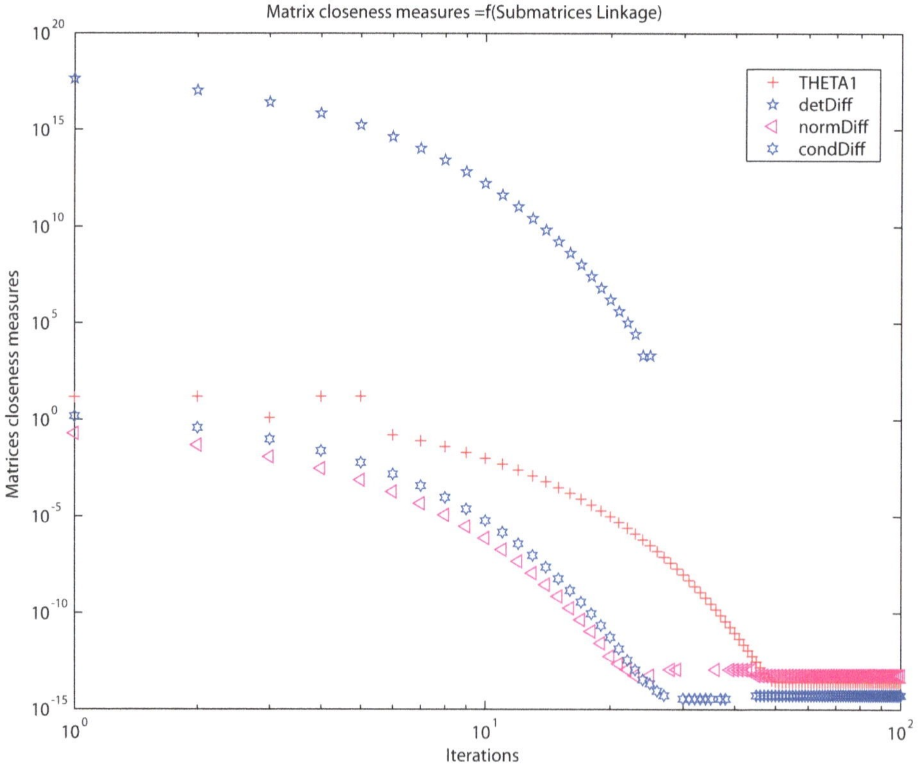

Figure 3.4.1. The results for Θ, $\Delta\det(\mathrm{R})$ and $\Delta\,\mathrm{cond}(\mathrm{R})$ criteria.

3.4.2: $rank(\mathbf{R}) = rank(\hat{\mathbf{R}}) = 10, rank(\mathbf{B}) = 2, rank(\mathbf{A}_1) = 2, rank(\mathbf{A}_2) = 8.$

Let $\hat{\mathbf{R}}$ will be 10-dimensional correlation matrix:

$$\hat{\mathbf{R}} = \begin{bmatrix} a_{11} & a_{12} & 0 & 0 & 0 & 0 & 0 & 0 & 0 & 0 \\ a_{21} & a_{22} & 0 & 0 & 0 & 0 & 0 & 0 & 0 & 0 \\ 0 & 0 & a_{33} & a_{34} & a_{35} & a_{36} & a_{37} & a_{38} & a_{39} & a_{310} \\ 0 & 0 & a_{43} & a_{44} & a_{45} & a_{46} & a_{47} & a_{48} & a_{49} & a_{410} \\ 0 & 0 & a_{53} & a_{54} & a_{55} & a_{56} & a_{57} & a_{58} & a_{59} & a_{510} \\ 0 & 0 & a_{63} & a_{64} & a_{65} & a_{66} & a_{67} & a_{68} & a_{69} & a_{610} \\ 0 & 0 & a_{73} & a_{74} & a_{75} & a_{76} & a_{77} & a_{78} & a_{79} & a_{710} \\ 0 & 0 & a_{83} & a_{84} & a_{85} & a_{86} & a_{87} & a_{88} & a_{89} & a_{810} \\ 0 & 0 & a_{93} & a_{94} & a_{95} & a_{96} & a_{97} & a_{98} & a_{99} & a_{910} \\ 0 & 0 & a_{103} & a_{104} & a_{105} & a_{106} & a_{107} & a_{108} & a_{109} & a_{1010} \end{bmatrix}$$

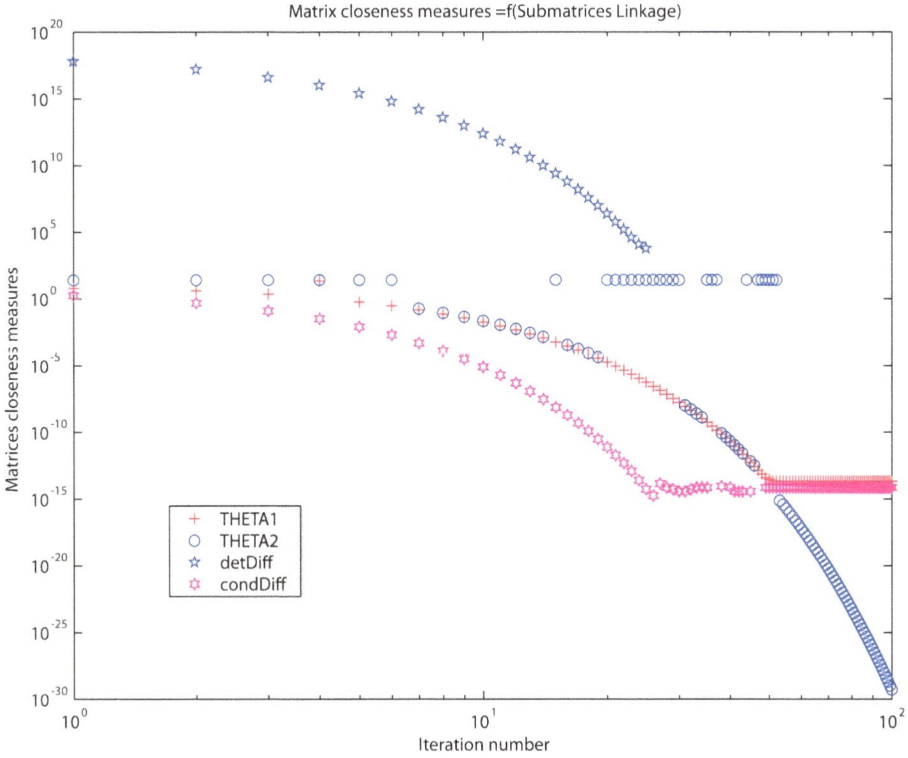

Figure 3.4.2. The results for Θ, $\Delta\,\det(\mathrm{R})$ and $\Delta\,\mathrm{cond}(\mathrm{R})$ criteria.

3.4.3: $rank(\mathbf{R}) = rank(\hat{\mathbf{R}}) = 10, rank(\mathbf{B}) = 3, rank(\mathbf{A}_1) = 3, rank(\mathbf{A}_2) = 7.$

Let $\hat{\mathbf{R}}$ will be 10-dimensional correlation matrix:

$$\hat{\mathbf{R}} = \begin{bmatrix} a_{11} & a_{12} & a_{13} & 0 & 0 & 0 & 0 & 0` & 0 & 0 \\ a_{21} & a_{22} & a_{23} & 0 & 0 & 0 & 0 & 0 & 0 & 0 \\ a_{31} & a_{32} & a_{33} & 0 & 0 & 0 & 0 & 0 & 0 & 0 \\ 0 & 0 & 0 & a_{44} & a_{45} & a_{46} & a_{47} & a_{48} & a_{49} & a_{410} \\ 0 & 0 & 0 & a_{54} & a_{55} & a_{56} & a_{57} & a_{58} & a_{59} & a_{510} \\ 0 & 0 & 0 & a_{64} & a_{65} & a_{66} & a_{67} & a_{68} & a_{69} & a_{610} \\ 0 & 0 & 0 & a_{74} & a_{75} & a_{76} & a_{77} & a_{78} & a_{79} & a_{710} \\ 0 & 0 & 0 & a_{84} & a_{85} & a_{86} & a_{87} & a_{88} & a_{89} & a_{810} \\ 0 & 0 & 0 & a_{94} & a_{95} & a_{96} & a_{97} & a_{98} & a_{99} & a_{910} \\ 0 & 0 & 0 & a_{104} & a_{105} & a_{106} & a_{107} & a_{108} & a_{109} & a_{1010} \end{bmatrix}$$

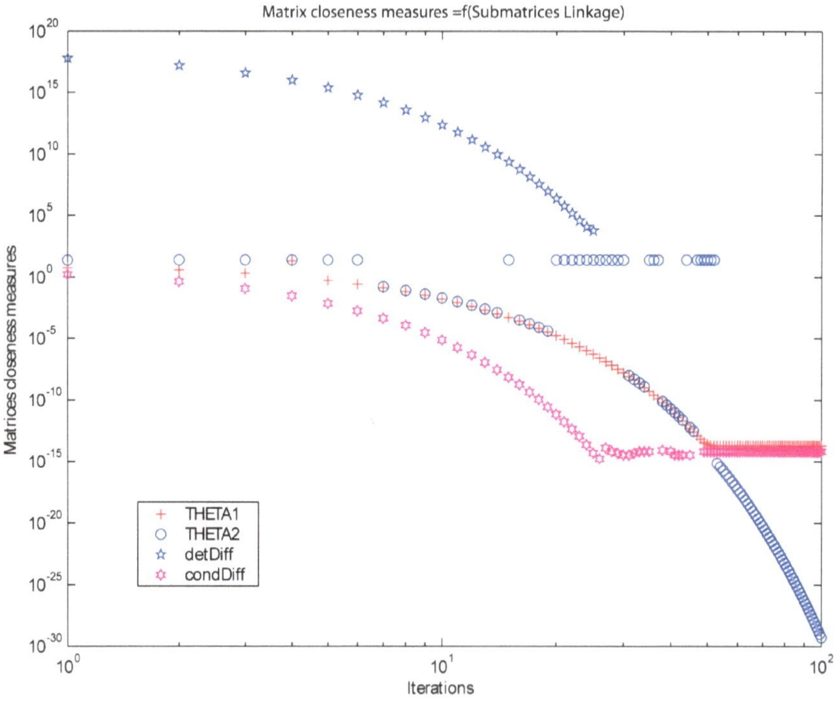

Figure 3.4.3. The results for Θ, $\Delta\det(\mathrm{R})$ and $\Delta\,\mathrm{cond}(\mathrm{R})$.

3.4.4.

$$rank(\mathbf{R}) = rank(\hat{\mathbf{R}}) = 10, rank(\mathbf{B}) = 4, rank(\mathbf{A}_1) = 4, rank(\mathbf{A}_2) = 6.$$

Let $\hat{\mathbf{R}}$ will be 10-dimensional correlation matrix:

$$\hat{\mathbf{R}} = \begin{bmatrix} a_{11} & a_{12} & a_{13} & a_{14} & 0 & 0 & 0 & 0 & 0 & 0 \\ a_{21} & a_{22} & a_{23} & a_{24} & 0 & 0 & 0 & 0 & 0 & 0 \\ a_{31} & a_{32} & a_{33} & a_{34} & 0 & 0 & 0 & 0 & 0 & 0 \\ a_{41} & a_{42} & a_{43} & a_{44} & 0 & 0 & 0 & 0 & 0 & 0 \\ 0 & 0 & 0 & 0 & a_{55} & a_{56} & a_{57} & a_{58} & a_{59} & a_{510} \\ 0 & 0 & 0 & 0 & a_{65} & a_{66} & a_{67} & a_{68} & a_{69} & a_{610} \\ 0 & 0 & 0 & 0 & a_{75} & a_{76} & a_{77} & a_{78} & a_{79} & a_{710} \\ 0 & 0 & 0 & 0 & a_{85} & a_{86} & a_{87} & a_{88} & a_{89} & a_{810} \\ 0 & 0 & 0 & 0 & a_{95} & a_{96} & a_{97} & a_{98} & a_{99} & a_{910} \\ 0 & 0 & 0 & 0 & a_{105} & a_{106} & a_{107} & a_{108} & a_{109} & a_{1010} \end{bmatrix}$$

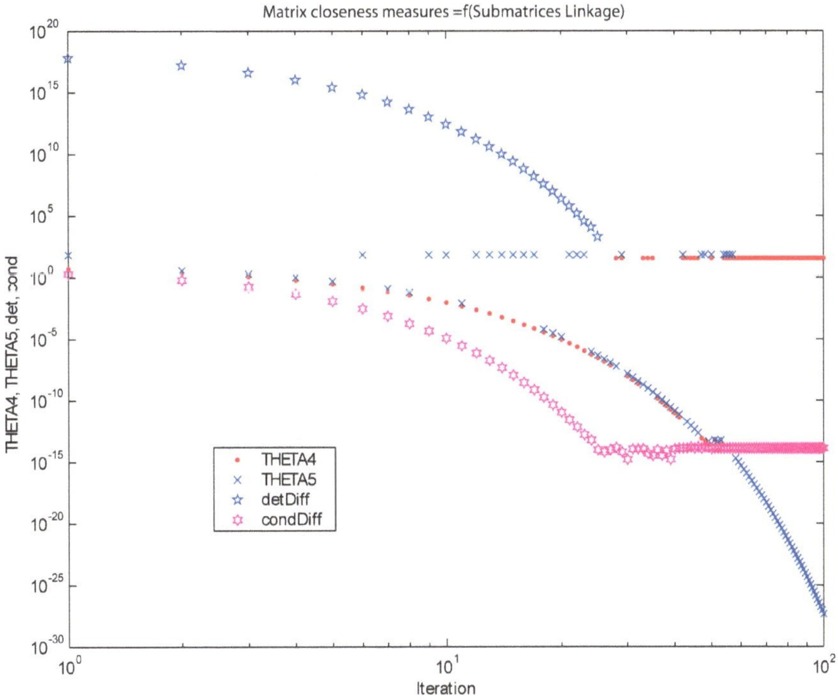

Figure 3.4.4. The results for Θ, $\Delta\,\det(\mathrm{R})$ and $\Delta\,\mathrm{cond}(\mathrm{R})$.

3.4.5. $rank(\mathbf{R}) = rank(\hat{\mathbf{R}}) = 10, rank(\mathbf{B}) = 5, rank(\mathbf{A}_1) = 5, rank(\mathbf{A}_2) = 5$.

Let $\hat{\mathbf{R}}$ will be 10-dimensional correlation matrix:

$$\hat{\mathbf{R}} = \begin{bmatrix} a_{11} & a_{12} & a_{13} & a_{14} & a_{15} & 0 & 0 & 0 & 0 & 0 \\ a_{21} & a_{22} & a_{23} & a_{24} & a_{25} & 0 & 0 & 0 & 0 & 0 \\ a_{31} & a_{32} & a_{33} & a_{34} & a_{35} & 0 & 0 & 0 & 0 & 0 \\ a_{41} & a_{42} & a_{43} & a_{44} & a_{45} & 0 & 0 & 0 & 0 & 0 \\ a_{51} & a_{52} & a_{53} & a_{54} & a_{55} & 0 & 0 & 0 & 0 & 0 \\ 0 & 0 & 0 & 0 & 0 & a_{66} & a_{67} & a_{68} & a_{69} & a_{610} \\ 0 & 0 & 0 & 0 & 0 & a_{76} & a_{77} & a_{78} & a_{79} & a_{710} \\ 0 & 0 & 0 & 0 & 0 & a_{86} & a_{87} & a_{88} & a_{89} & a_{810} \\ 0 & 0 & 0 & 0 & 0 & a_{96} & a_{97} & a_{98} & a_{99} & a_{910} \\ 0 & 0 & 0 & 0 & 0 & a_{106} & a_{107} & a_{108} & a_{109} & a_{1010} \end{bmatrix}$$

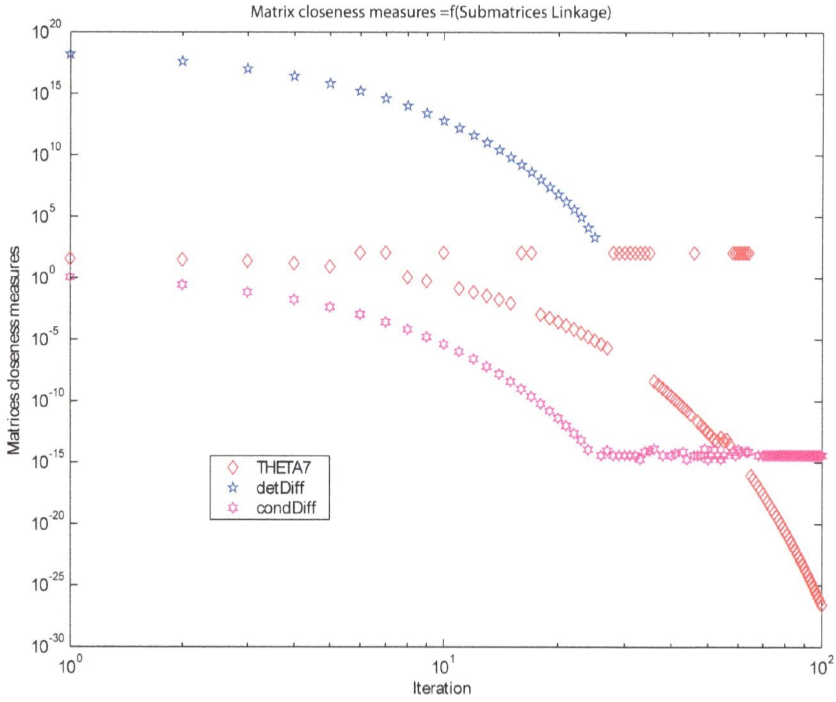

Figure 3.4.5. The results for Θ, $\Delta\,\det(\mathrm{R})$ and $\Delta\,\mathrm{cond}(\mathrm{R})$.

3.4.6. $rank(\mathbf{R}) = rank(\hat{\mathbf{R}}) = 10, rank(\mathbf{B}) = 4, rank(\mathbf{A}_1) = 6, rank(\mathbf{A}_2) = 4$.

Let $\hat{\mathbf{R}}$ will be 10-dimensional correlation matrix:

$$\hat{\mathbf{R}} = \begin{bmatrix} a_{11} & a_{12} & a_{13} & a_{14} & a_{15} & a_{16} & 0 & 0 & 0 & 0 \\ a_{21} & a_{22} & a_{23} & a_{24} & a_{25} & a_{26} & 0 & 0 & 0 & 0 \\ a_{31} & a_{32} & a_{33} & a_{34} & a_{35} & a_{36} & 0 & 0 & 0 & 0 \\ a_{41} & a_{42} & a_{43} & a_{44} & a_{45} & a_{46} & 0 & 0 & 0 & 0 \\ a_{51} & a_{52} & a_{53} & a_{54} & a_{55} & a_{56} & 0 & 0 & 0 & 0 \\ a_{61} & a_{62} & a_{63} & a_{64} & a_{65} & a_{66} & 0 & 0 & 0 & 0 \\ 0 & 0 & 0 & 0 & 0 & 0 & a_{77} & a_{78} & a_{79} & a_{710} \\ 0 & 0 & 0 & 0 & 0 & 0 & a_{87} & a_{88} & a_{89} & a_{810} \\ 0 & 0 & 0 & 0 & 0 & 0 & a_{97} & a_{98} & a_{99} & a_{910} \\ 0 & 0 & 0 & 0 & 0 & 0 & a_{107} & a_{108} & a_{109} & a_{1010} \end{bmatrix}$$

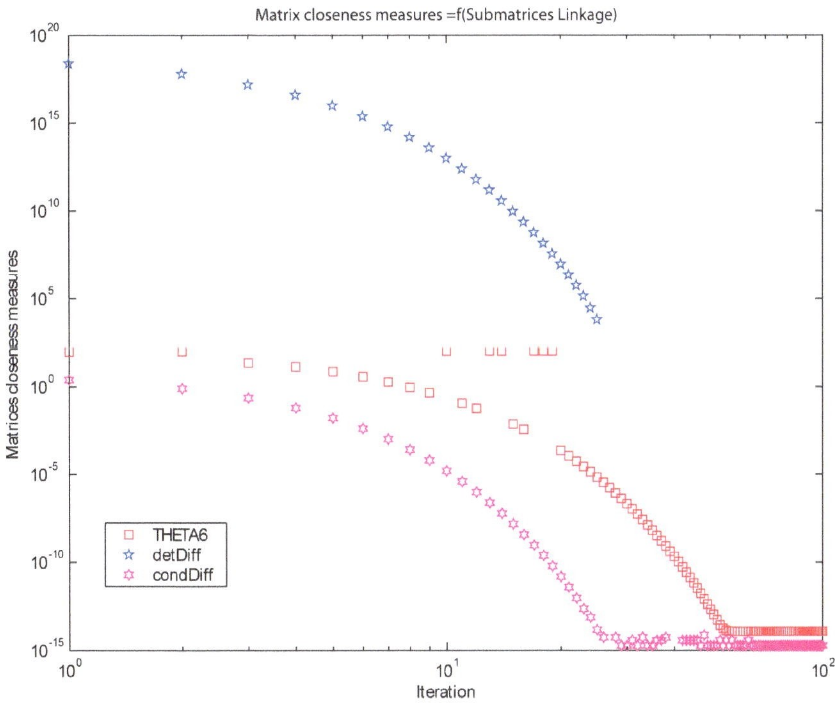

Figure 3.4.6. The results for Θ, $\Delta\,\det(\mathrm{R})$ and $\Delta\,\mathrm{cond}(\mathrm{R})$.

3.4.7: $rank(\mathbf{R}) = rank(\hat{\mathbf{R}}) = 10, rank(\mathbf{B}) = 3, rank(\mathbf{A}_1) = 7, rank(\mathbf{A}_2) = 3$.

Let $\hat{\mathbf{R}}$ will be 10-dimensional correlation matrix:

$$\hat{\mathbf{R}} = \begin{bmatrix} a_{11} & a_{12} & a_{13} & a_{14} & a_{15} & a_{16} & a_{17} & 0 & 0 & 0 \\ a_{21} & a_{22} & a_{23} & a_{24} & a_{25} & a_{26} & a_{27} & 0 & 0 & 0 \\ a_{31} & a_{32} & a_{33} & a_{34} & a_{35} & a_{36} & a_{37} & 0 & 0 & 0 \\ a_{41} & a_{42} & a_{43} & a_{44} & a_{45} & a_{46} & a_{47} & 0 & 0 & 0 \\ a_{51} & a_{52} & a_{53} & a_{54} & a_{55} & a_{56} & a_{57} & 0 & 0 & 0 \\ a_{61} & a_{62} & a_{63} & a_{64} & a_{65} & a_{66} & a_{67} & 0 & 0 & 0 \\ a_{71} & a_{72} & a_{73} & a_{74} & a_{75} & a_{76} & a_{77} & 0 & 0 & 0 \\ 0 & 0 & 0 & 0 & 0 & 0 & 0 & a_{88} & a_{89} & a_{810} \\ 0 & 0 & 0 & 0 & 0 & 0 & 0 & a_{98} & a_{99} & a_{910} \\ 0 & 0 & 0 & 0 & 0 & 0 & 0 & a_{108} & a_{109} & a_{1010} \end{bmatrix}$$

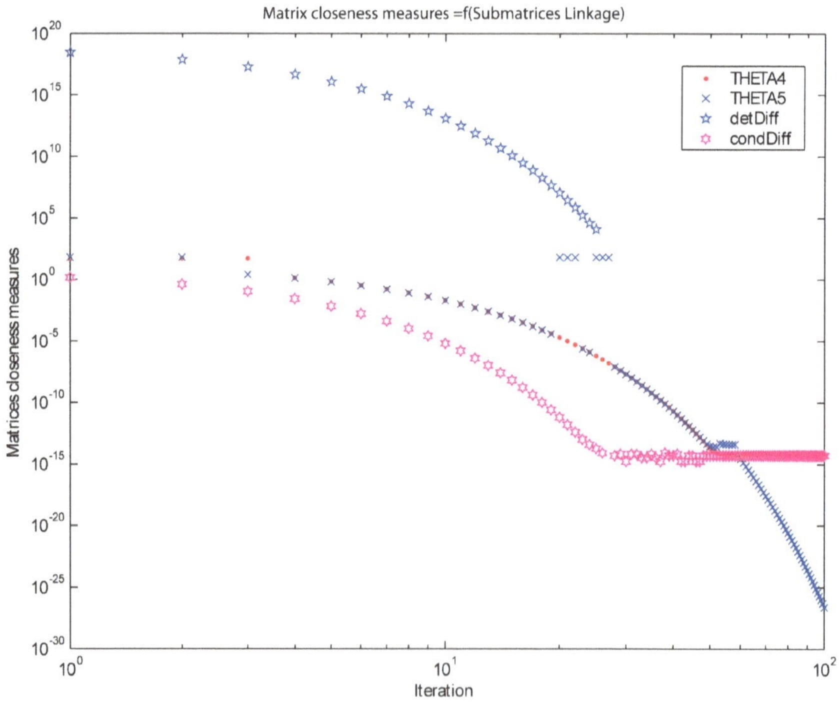

Figure 3.4.7. The results for Θ, $\Delta\det(\mathrm{R})$ and $\Delta\,\mathrm{cond}(\mathrm{R})$.

3.4.8. $rank(\mathbf{R}) = rank(\hat{\mathbf{R}}) = 10, rank(\mathbf{B}) = 2, rank(\mathbf{A}_1) = 8, rank(\mathbf{A}_2) = 2$.

Let $\hat{\mathbf{R}}$ will be 10-dimensional correlation matrix:

$$\hat{\mathbf{R}} = \begin{bmatrix} a_{11} & a_{12} & a_{13} & a_{14} & a_{15} & a_{16} & a_{17} & a_{18} & 0 & 0 \\ a_{21} & a_{22} & a_{23} & a_{24} & a_{25} & a_{26} & a_{27} & a_{28} & 0 & 0 \\ a_{31} & a_{32} & a_{33} & a_{34} & a_{35} & a_{36} & a_{37} & a_{38} & 0 & 0 \\ a_{41} & a_{42} & a_{43} & a_{44} & a_{45} & a_{46} & a_{47} & a_{48} & 0 & 0 \\ a_{51} & a_{52} & a_{53} & a_{54} & a_{55} & a_{56} & a_{57} & a_{58} & 0 & 0 \\ a_{61} & a_{62} & a_{63} & a_{64} & a_{65} & a_{66} & a_{67} & a_{68} & 0 & 0 \\ a_{71} & a_{72} & a_{73} & a_{74} & a_{75} & a_{76} & a_{77} & a_{78} & 0 & 0 \\ a_{81} & a_{82} & a_{83} & a_{84} & a_{85} & a_{86} & a_{87} & a_{88} & 0 & 0 \\ 0 & 0 & 0 & 0 & 0 & 0 & 0 & 0 & a_{99} & a_{910} \\ 0 & 0 & 0 & 0 & 0 & 0 & 0 & 0 & a_{109} & a_{1010} \end{bmatrix}$$

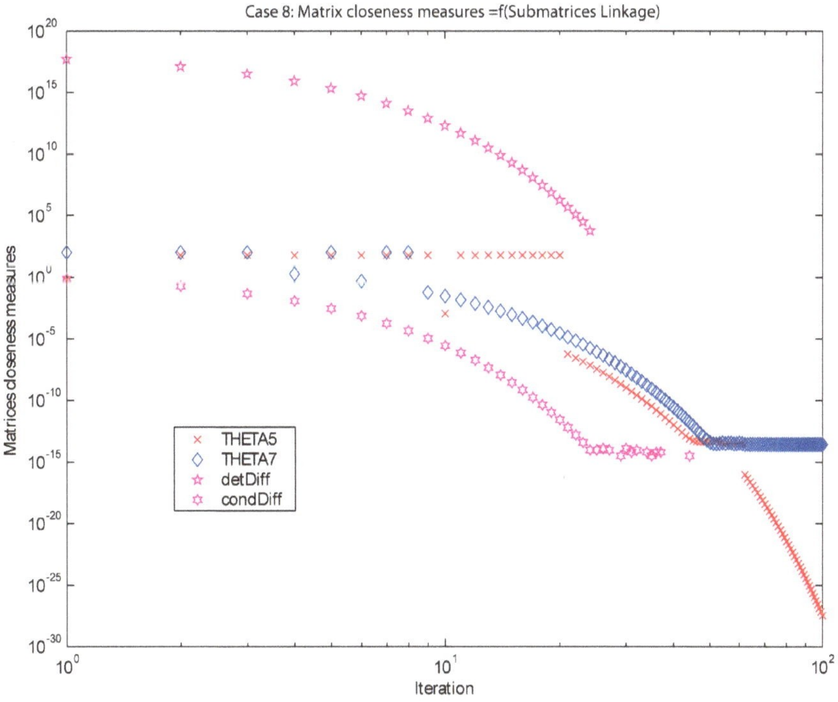

Figure 3.4.8. The results for Θ, $\Delta\,\det(\mathrm{R})$ and $\Delta\,\mathrm{cond}(\mathrm{R})$.

3.4.9: $rank(\mathbf{R}) = rank(\hat{\mathbf{R}}) = 10, rank(\mathbf{B}) = 1, rank(\mathbf{A}_1) = 9, rank(\mathbf{A}_2) = 1$.

Let $\hat{\mathbf{R}}$ will be 10-dimensional correlation matrix:

$$\hat{\mathbf{R}} = \begin{bmatrix} a_{11} & a_{12} & a_{13} & a_{14} & a_{15} & a_{16} & a_{17} & a_{18} & a_{19} & 0 \\ a_{21} & a_{22} & a_{23} & a_{24} & a_{25} & a_{26} & a_{27} & a_{28} & a_{29} & 0 \\ a_{31} & a_{32} & a_{33} & a_{34} & a_{35} & a_{36} & a_{37} & a_{38} & a_{39} & 0 \\ a_{41} & a_{42} & a_{43} & a_{44} & a_{45} & a_{46} & a_{47} & a_{48} & a_{49} & 0 \\ a_{51} & a_{52} & a_{53} & a_{54} & a_{55} & a_{56} & a_{57} & a_{58} & a_{59} & 0 \\ a_{61} & a_{62} & a_{63} & a_{64} & a_{65} & a_{66} & a_{67} & a_{68} & a_{69} & 0 \\ a_{71} & a_{72} & a_{73} & a_{74} & a_{75} & a_{76} & a_{77} & a_{78} & a_{79} & 0 \\ a_{81} & a_{82} & a_{83} & a_{84} & a_{85} & a_{86} & a_{87} & a_{88} & a_{89} & 0 \\ a_{91} & a_{92} & a_{93} & a_{94} & a_{95} & a_{96} & a_{97} & a_{98} & a_{99} & 0 \\ 0 & 0 & 0 & 0 & 0 & 0 & 0 & 0 & 0 & a_{1010} \end{bmatrix}$$

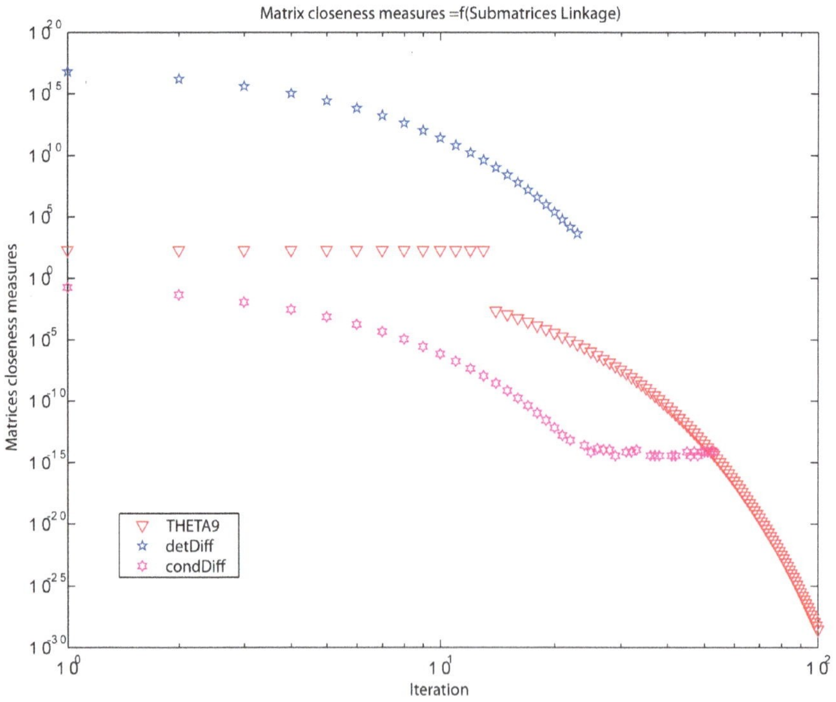

Figure 3.4.9. The results for Θ, $\Delta\,\det(\mathrm{R})$ and $\Delta\,\mathrm{cond}(\mathrm{R})$.

Table 1.

Criteria Type	Interval	Criteria graphical description
3.2. Sequence *R* of matrices $\mathbf{R}_i, rank(\mathbf{R}_i) = 4$		
3.2.1. $\text{rank}\,\mathbf{R} = 4; \text{rank}\,\mathbf{R}_1 = \text{rank}\,\mathbf{R}_2 = 2$		
$\Theta_1(\mathbf{R}, \hat{\mathbf{R}})$	(1, 100)	(1, 100)- Convex decreasing curve;
$\Theta_2(\mathbf{R}, \hat{\mathbf{R}})$	(1, 100)	(1, 100)- Convex decreasing curve;
$\Delta \det(\mathbf{R}, \hat{\mathbf{R}})$	(1, 12)	(1, 12) - Segment of Convex decreasing curve; (12, 100) - Segment of Horizontal line
$\Delta \text{cond}(\mathbf{R}, \hat{\mathbf{R}})$	(1, 12)	(1, 12) - Segment of Convex decreasing curve; (12, 100) - Segment of Horizontal line
3.3. Sequence *R* Matrices $\mathbf{R}_i, rank(\mathbf{R}_i) = 5$		
3.3.1. $\text{rank}(\mathbf{R}) = \text{rank}(\hat{\mathbf{R}}) = 5, \text{rank}(\mathbf{B}_i) = \text{rank}(\mathbf{A}_1) = 2, \text{rank}(\mathbf{A}_2) = 3$.		
$\Theta_1(\mathbf{R}, \hat{\mathbf{R}})$	[1,20], [25,100]	2 segments of convex decreasing curve with small horizontal plateau (20,25) in the interval
$\Theta_2(\mathbf{R}, \hat{\mathbf{R}})$	[1,23], [25,100]	2 segments of convex decreasing curve with small horizontal plateau (23,25) in the interval
$\Delta \det(\mathbf{R}, \hat{\mathbf{R}})$	(1, 12)	(1, 12) - Segment of Convex decreasing curve; (12, 100) - Segment of Horizontal line
$\Delta \text{cond}(\mathbf{R}, \hat{\mathbf{R}})$	(1, 12)	(1, 12) - Segment of Convex decreasing curve; (12, 100) - Segment of Horizontal line
3.3.2: $\text{rank}(\mathbf{R}) = \text{rank}(\hat{\mathbf{R}}) = 5, \text{rank}(\mathbf{B}_i) = 2, \text{rank}(\mathbf{A}_1) = 3, \text{rank}(\mathbf{A}_2) = 2$		
$\Theta_2(\mathbf{R}, \hat{\mathbf{R}})$	(1, 100)	(1, 100) - Convex decreasing curve
$\Theta_5(\mathbf{R}, \hat{\mathbf{R}})$	(1,20)	(1, 20) - Segment of Convex decreasing curve; (20, 100) - Segment of Horizontal line
$\Delta \det(\mathbf{R}, \hat{\mathbf{R}})$	(1, 12)	(1, 12) - Segment of Convex decreasing curve; (12, 100) - Segment of Horizontal line
$\Delta \text{cond}(\mathbf{R}, \hat{\mathbf{R}})$	(1, 12)	(1, 12) - Segment of Convex decreasing curve; (12, 100) - Segment of Horizontal line

3.4. Sequence *R* Matrices $\mathbf{R}_i, rank(\mathbf{R}_i) = 10$

3.4.1: $\text{rank}\,\mathbf{R} = 10; \text{rank}\,\mathbf{R}_1 = 1; \text{rank}\,\mathbf{R}_2 = 9$		
$\Theta_1(\mathbf{R},\hat{\mathbf{R}})$	(1,50)	(1,50) - Convex decreasing curve; (50, 100) -Horizontal line
$\Theta_2(\mathbf{R},\hat{\mathbf{R}})$	(1,25)	(1,25) - Segment of Convex decreasing curve; (25,100) - Segment of Horizontal line
$\Delta\det(\mathbf{R},\hat{\mathbf{R}})$	(1,25)	(1,25) - Segment of Convex decreasing curve; (25,100) - Segment of Horizontal line
$\Delta\text{cond}(\mathbf{R},\hat{\mathbf{R}})$	(1,25)	(1,25) - Segment of Convex decreasing curve; (25,100) - Segment of Horizontal line
3.4.2: $\text{rank}\,\mathbf{R} = 10; \text{rank}\,\mathbf{R}_1 = 2; \text{rank}\,\mathbf{R}_2 = 8$		
$\Theta_1(\mathbf{R},\hat{\mathbf{R}})$	(1,50)	(1,50) - Convex decreasing curve; (50, 100)-Horizontal line
$\Theta_2(\mathbf{R},\hat{\mathbf{R}})$	(6,20), (50, 100)	(1,5) - Segment of horizontal line; (6,20) - Segment of Convex decreasing curve; (20,50) - Segment of horizontal line; (50,100) - Segment of Convex decreasing curve
$\Delta\det(\mathbf{R},\hat{\mathbf{R}})$	(1,25)	(1,25) - Segment of Convex decreasing curve; (25,100) - Segment of Horizontal line
$\Delta\text{cond}(\mathbf{R},\hat{\mathbf{R}})$	(1,25)	(1,25) - Segment of Convex decreasing curve; (25,100) - Segment of Horizontal line
3.4.3: $\text{rank}\,\mathbf{R} = 10; \text{rank}\,\mathbf{R}_1 = 3; \text{rank}\,\mathbf{R}_2 = 7$		
$\Theta_4(\mathbf{R},\hat{\mathbf{R}})$	(1,40)	(1,40) - Convex decreasing curve with 2 small segments of horizontal line; (40, 100) - Horizontal line
$\Theta_5(\mathbf{R},\hat{\mathbf{R}})$	(20, 100)	(1, 20) - Segment of horizontal line; (20,100) - Convex decreasing curve with 2 small segments of horizontal line;

$\Delta \det(\mathbf{R}, \hat{\mathbf{R}})$	(1,25)	(1,25) - Segment of Convex decreasing curve; (25,100) - Segment of Horizontal line
$\Delta \mathrm{cond}(\mathbf{R}, \hat{\mathbf{R}})$	(1,25)	(1,25) - Segment of Convex decreasing curve; (25,100) - Segment of Horizontal line

3.4.4: $\mathrm{rank}\mathbf{R} = 10; \mathrm{rank}\mathbf{R}_1 = 4; \mathrm{rank}\mathbf{R}_2 = 6$

$\Theta_1(\mathbf{R}, \hat{\mathbf{R}})$	(10, 100)	(10,100) - Segment of Convex decreasing curve with 4 small segments of horizontal line; (40, 100) -Horizontal line
$\Theta_5(\mathbf{R}, \hat{\mathbf{R}})$	(1,50)	(1,50) - Segment of Convex decreasing curve; (50, 100) - Segment of horizontal line;
$\Delta \det(\mathbf{R}, \hat{\mathbf{R}})$	(1,25)	(1,25) - Segment of Convex decreasing curve; (25,100) - Segment of Horizontal line
$\Delta \mathrm{cond}(\mathbf{R}, \hat{\mathbf{R}})$	(1,25)	(1,25) - Segment of Convex decreasing curve; (25,100) - Segment of Horizontal line

3.4.5: $\mathrm{rank}\mathbf{R} = 10; \mathrm{rank}\mathbf{R}_1 = 5; \mathrm{rank}\mathbf{R}_2 = 5$

$\Theta_7(\mathbf{R}, \hat{\mathbf{R}})$	(1,100)	(1,100) - Segment of Convex decreasing curve with 4 small segments of horizontal line
$\Delta \det(\mathbf{R}, \hat{\mathbf{R}})$	(1,25)	(1,25) - Segment of Convex decreasing curve; (25,100) - Segment of Horizontal line
$\Delta \mathrm{cond}(\mathbf{R}, \hat{\mathbf{R}})$	(1,25)	(1,25) - Segment of Convex decreasing curve; (25,100) - Segment of Horizontal line

3.4.6: $\mathrm{rank}\mathbf{R} = 10; \mathrm{rank}\mathbf{R}_1 = 6; \mathrm{rank}\mathbf{R}_2 = 4$

$\Theta_6(\mathbf{R}, \hat{\mathbf{R}})$	(1,55)	(1,55) - Segment of Convex decreasing curve with 2 small segments of horizontal line; (55,100) - Segment of Horizontal line

$\Delta\det\left(\mathbf{R},\hat{\mathbf{R}}\right)$	(1,25)	(1,25) - Segment of Convex decreasing curve; (25,100) - Segment of Horizontal line
$\Delta\mathrm{cond}\left(\mathbf{R},\hat{\mathbf{R}}\right)$	(1,25)	(1,25) - Segment of Convex decreasing curve; (25,100) - Segment of Horizontal line
3.4.7: $\mathrm{rank}\mathbf{R}=10;\mathrm{rank}\mathbf{R}_1=7;\mathrm{rank}\mathbf{R}_2=3\}$		
$\Theta_1\left(\mathbf{R},\hat{\mathbf{R}}\right)$	(1,50)	(1,50) - Segment of Convex decreasing curve with 2 small segments of horizontal line; (50,100) - Segment of Horizontal line
$\Delta\det\left(\mathbf{R},\hat{\mathbf{R}}\right)$	(1,25)	(1,25) - Segment of Convex decreasing curve; (1,100) - Segment of Horizontal line
$\Delta\mathrm{cond}\left(\mathbf{R},\hat{\mathbf{R}}\right)$	(1,25)	(1,25) - Segment of Convex decreasing curve; (1,100) - Segment of Horizontal line
3.4.8: $\mathrm{rank}\mathbf{R}=10;\mathrm{rank}\mathbf{R}_1=8;\mathrm{rank}\mathbf{R}_2=2$		
$\Theta_5\left(\mathbf{R},\hat{\mathbf{R}}\right)$	(20,45), (60,100)	(1,20) - Segment of Horizontal line; (20,45) - Segment of Convex decreasing curve; (45,60) Segment of horizontal line; (60,100) - Segment of Convex decreasing curve
$\Theta_7\left(\mathbf{R},\hat{\mathbf{R}}\right)$	(1,50)	(1,50) - Segment of Convex decreasing curve with 3 "wild points"; (50,100) - Segment of Horizontal line
$\Delta\det\left(\mathbf{R},\hat{\mathbf{R}}\right)$	(1,25)	(1,25) - Segment of Convex decreasing curve; (1,100) -

		Segment of Horizontal line
$\Delta\,\mathrm{cond}\left(\mathbf{R},\hat{\mathbf{R}}\right)$	(1,25)	(1,25) - Segment of Convex decreasing curve; (1,100) - Segment of Horizontal line
3.4.9 $\mathrm{rank}\,\mathbf{R}=10;\mathrm{rank}\,\mathbf{R}_1=9;\mathrm{rank}\,\mathbf{R}_2=1$		
$\Theta_9\left(\mathbf{R},\hat{\mathbf{R}}\right)$	(15,100)	(1,15) - Segment of Horizontal line; (15,100) - Segment of Convex decreasing curve
$\Delta\det\left(\mathbf{R},\hat{\mathbf{R}}\right)$	(1,25)	(1,25) - Segment of Convex decreasing curve; (1,100) - Segment of Horizontal line
$\Delta\,\mathrm{cond}\left(\mathbf{R},\hat{\mathbf{R}}\right)$	(1,25)	(1,25) - Segment of Convex decreasing curve; (1,100) - Segment of Horizontal line

3.5. The matrices block linkage $LINK$ and $\det\hat{R}$ are variable

Let's construct two sequences of matrices:

$\Re = \{R_{ik}\}_{11}^{NK}$ with $\det R_{ik_1} \lhd \det R_{ik_2}$, if $k_1 \rhd k_2$, $\forall k_1, k_2 \in 1, K$,

$\mathbf{B}_i = LINK^i * \mathbf{B}$, $i = 1, N$; $0 < LINK < 1$. Then

$\lim_{i\to\infty} \mathbf{B}_i = \mathbf{0}$ and $\lim_{i\to\infty} \Theta(\mathbf{R}_i, \hat{\mathbf{R}}) = 0.$.

and

$\hat{\Re} = \{\hat{R}_k\}_1^K$

where

$$R_{ik} = \begin{bmatrix} A_{1k} & B_{ik}^T \\ B_{ik} & A_2 \end{bmatrix}, \quad \hat{R}_k = \begin{bmatrix} A_{1k} & 0 \\ 0 & A_2 \end{bmatrix}$$

with $\det R_{k_1} \lhd \det R_{k_2}$ and $\det\hat{R}_{k_1} \lhd \det\hat{R}_{k_2}$, if $k_1 \rhd k_2$, $\forall k_1, k_2 \in 1, K$.

Let construct the response functions $\mathbf{F}_\Theta$, $\mathbf{F}_{\Delta\det\mathbf{R}}$ and $\mathbf{F}_{\Delta\mathrm{cond}\,\mathbf{R}}$ on surface

$(LINK, \ \mathbf{det}\ \hat{R})$ for $\Theta_1(\mathbf{R}_i, \hat{\mathbf{R}})$, $\Delta\mathbf{det}\ R$ and $\Delta\mathbf{cond}\ R$.

$\mathbf{F}_\Theta = \mathbf{F}(LINK, \ \mathbf{det}\ \hat{R}, \Theta_1(\mathbf{R}_i, \hat{\mathbf{R}}))$, $LINK \in \{10^{-16}, 10^{0}\}$, $\mathbf{det}\ \hat{R} \in$ $\{10^{7}, 10^{13}\}$.

The $\mathbf{F}_{\Theta_1(\mathbf{R},\hat{\mathbf{R}})}$ presented as contours **on** Fig. 4, $\mathbf{F}_{\Delta\det\mathbf{R}}$ - **on** Fig. 5 and

$\mathbf{F}_{\Delta\mathrm{cond}\,\mathbf{R}}$ - on Fig. 6.

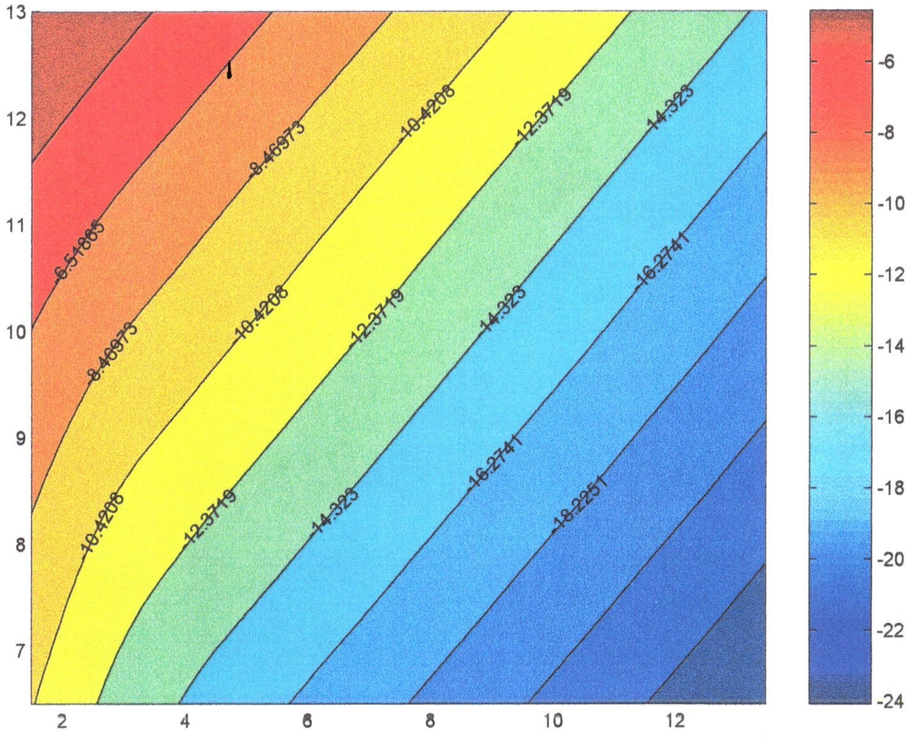

Figure 4.

Response function $F_{\theta} = F(\ LINK, \det \hat{\mathbf{R}}, \Theta)$,

$LINK \in (10^{-16}, 10^{0}), \det \hat{\mathbf{R}} \in (10^{7}, 10^{13})$

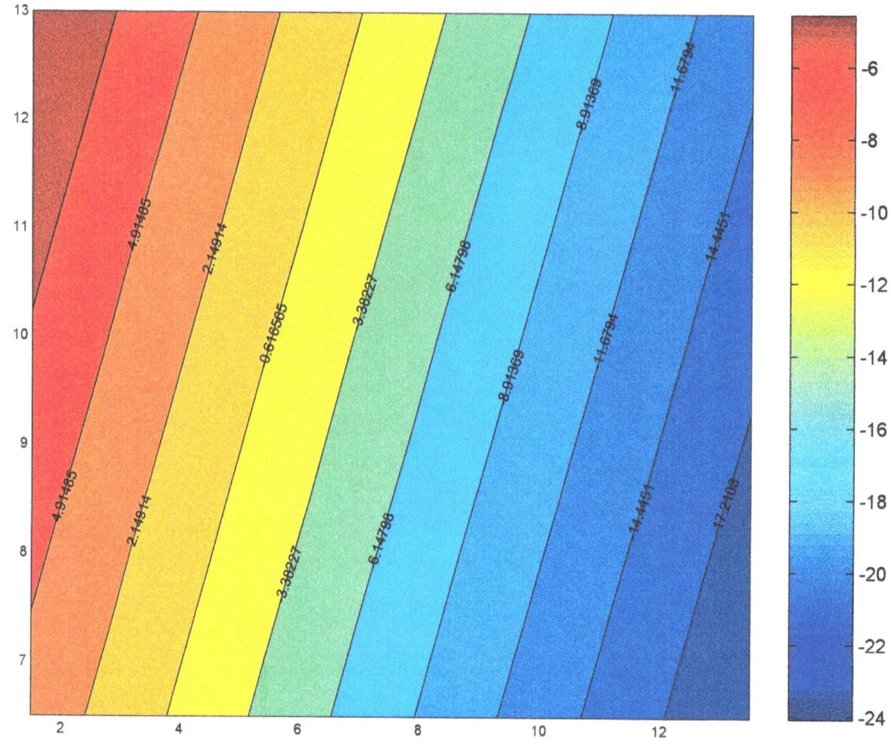

Figure 5.

Response function $F_{|\det R - \det \hat{\mathbf{R}}|} = F(\ LINK, \det \hat{\mathbf{R}}, |\det R - \det \hat{\mathbf{R}}|),$

$LINK \in (10^{-16}, 10^{0}), \det \hat{\mathbf{R}} \in (10^{7}, 10^{13})$

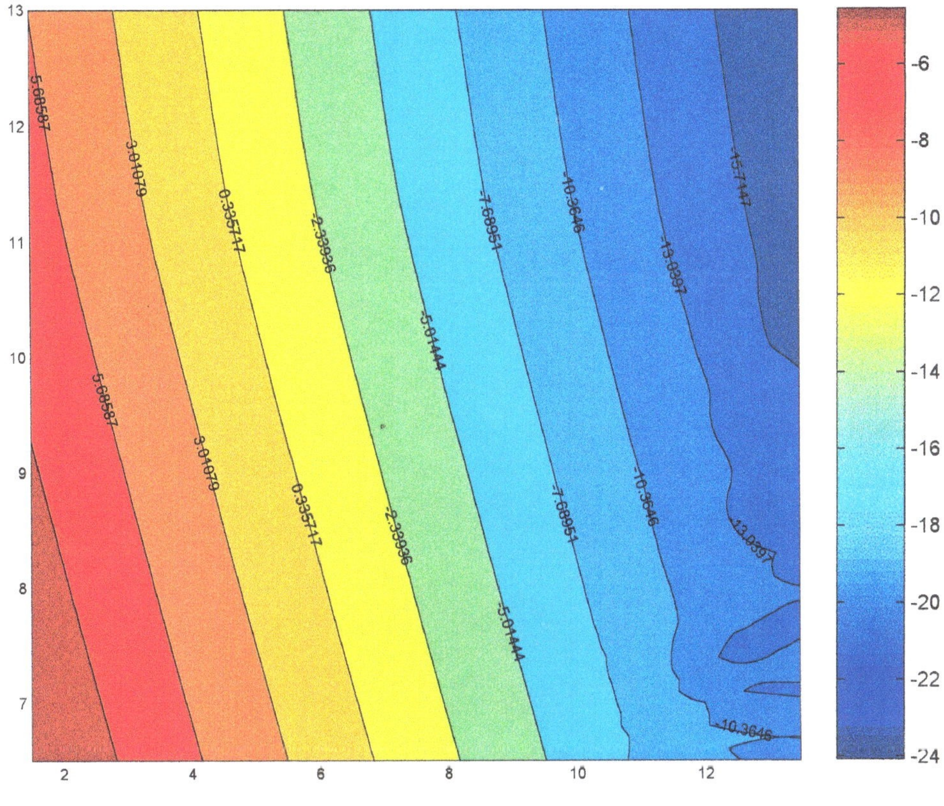

Figure 6.

Response function $F_{|condR - cond\hat{R}|} = F(\ LINK, \det \hat{\mathbf{R}}, \left|condR - cond\hat{\mathbf{R}}\right|)$,

$LINK \in \left(10^{-16}, 10^{0}\right), \det \hat{\mathbf{R}} \in \left(10^{7}, 10^{13}\right)$

Table 2.

Criteria Type	*LINK* Correct Region	Det R Correct Region	Incorrect Region for *LINK*	Det R Incorrect Region	Criteria Slope
$\Theta_1(\mathbf{R}_i, \hat{\mathbf{R}})$	(-14, -3)	(6, 13)	None	None	1
$\Delta \det R$	(-14, -1.5)	(6, 13)	None	None	2
Δcond R	(-10, -1.5)	(8, 13)	(-13, -10)	(6.5, 10)	2

Conclusions for Numerical Experiments

1. For det R_1 = const and $LINK$ = Var:
   - Criteria $\Theta_1(\mathbf{R}_i, \hat{\mathbf{R}})$ correctly represented matrices linkage decrease process for all iterations for **3.2.1, 3.3.1,** Cases.
   - Criteria $\Theta_2(\mathbf{R}_i, \hat{\mathbf{R}})$ correctly represented matrices linkage decrease process for all iterations for **3.3.1, 3.3.2** Cases.
   - Criteria $\Delta(\det(\mathbf{R}))$, $\Delta(\text{cond}(\mathbf{R}))$ adequately represented process only for 10 first iterations for all **3.4.** Cases.
   - There are several Theta Criteria, applicable for **3.4.** Cases. They are better in accuracy than $\Delta(\det(\mathbf{R}))$, $\Delta(\text{cond}(\mathbf{R}))$.
   - In all of our experiments exists at least one Θ criteria with superior accuracy to $\Delta(\det(\mathbf{R}))$ and $\Delta(\text{cond}(\mathbf{R}))$ for identification of very weak linkages between matrices.

2. For det R_1 = Var, $LINK$ = Var :
   a. The criteria $\Theta_1(\mathbf{R}_i, \hat{\mathbf{R}})$ is correct for the whole domain with constant slope m = 1.
   b. The criteria $\Delta\mathbf{det}\, R$ is very steep and can be used for the whole domain.
   c. The criteria $\Delta\mathbf{cond}\, R$ was incorrect for small $LINK$.
   d. The Θ criteria have a tremendous potential for improving accuracy in analysis of multi-dimensional objects and systems.

References:

1. G. Golub, C. Van Loan. Matrix Computations. The John Hopkins University Press, 1996.
2. C. De Boor. A Practical Guide to Splines. Springer-Verlag, 1978.

CHAPTER IV

Applications

4.1. Preliminary Conditions

Let us specify preliminary conditions for Theta Criteria usage:

- Standard Statistical Regression Analysis has been performed.
- Accuracy of Standard Statistical Regression Analysis is inadequate.
- Application can be described by set of multiple variables.
- The application data accuracy is suitable for evaluation of its spectral characteristics.
- Available statistical or mathematical software tool to evaluate eigenvalues and eigenvectors.

We will now discuss various application areas for Theta Criteria.

4.2. IED Identification

Improvised Explosive Devices (IED) are made from 5 basic types of plastic explosives: C-4, PENO, Primasheet, RDX and Semtex. The IED can also be made using over-the-counter chemicals: aspirin, phenol, bleach, pool chorine compound, etc. These explosives are concealed underground, inside metal structures or strapped to human body.

Regrettably, existing identification methods do not have the desired accuracy to detect IED. We proposed our recommendation for IED identification to U.S. Department of Defense (DoD).

4.3. Aircraft Engine Failure Identification

Theta Criteria can be used for Aircraft Engine Failure Identification. Implementation details are available upon request.

4.4. Medical Applications

Theta Criteria can be used for identification and therapy of various diseases, disorders and illnesses. Implementation details are available upon request.

4.5. Financial Problems Analysis

Theta Criteria can be used for Financial Application Analysis. Implementation details are available upon request.

INDEX

www.ingramcontent.com/pod-product-compliance
Lightning Source LLC
Chambersburg PA
CBHW050755180526
45159CB00003B/1473
* 9 7 8 1 4 5 6 4 6 7 2 0 3 *